CIRIA C652

London, 2006

Safer surfaces to walk on

reducing the risk of slipping

CIRIA *sharing knowledge* ■ *building best practice*

Classic House, 174–180 Old Street, London EC1V 9BP, UK
TEL: +44 (0)20 7549 3300   FAX +44 (0)20 7253 0523
EMAIL enquiries@ciria.org
WEBSITE www.ciria.org

Safer surfaces to walk on – reducing the risk of slipping.

Carpenter, J; Lazarus, D; Perkins, C

CIRIA

CIRIA C652   © CIRIA 2006   RP713   ISBN-13: 978-0-86017-652-7

ISBN-10: 0-86017-652-5

British Library Cataloguing in Publication Data

A catalogue record for this book is available from the British Library.

Keywords
Health and safety, facilities management, materials technology, refurbishment, regulation, respect for people, risk and value management, transport infrastructure

Reader interest	Classification	
Health and safety, risk assessment, materials, materials technology, design, specification, facilities management, maintenance, flooring products	AVAILABILITY	Unrestricted
	CONTENT	Advice/guidance
	STATUS	Committee-guided
	USER	Building owners and operators, construction industry designers and specifiers, architects, engineers, surveyors, contractors and consultants, facilities managers and maintenance contractors, manufacturers, suppliers and installers of flooring products and of floor cleaning products and equipment, health and safety professionals, local authorities

Published by CIRIA, Classic House, 174–180 Old Street, London EC1V 9BP, UK.

Foreword

Slipping accidents cause many thousands of occupational major injuries each year, more than 90 per cent of which involve broken bones, and much pain, suffering and financial loss for society. Yet the perception by the public, the workforce and those who design or manage floors in buildings is that these accidents are inevitable. The reality is that sensible precautions could eliminate the majority of these accidents. As with many problems, we believe that if the problem is explained, and sensible solutions are suggested, then people will start to manage the risks more effectively.

This CIRIA good practice guide has been produced as part of the Health and Safety Commission's programme to reduce the number of slips and trips accidents in the UK. It is an attempt to give practical guidance to those who design, procure and manage flooring in buildings and is based upon many research projects undertaken, primarily by the Health and Safety Laboratory.

These research projects looked at the science and mechanics of why people slip. The Health and Safety Executive, armed with the information from this research, have looked at the other contributing factors that may initiate a slip – flooring material, contaminant, shoe soles, cleaning regimes and environment issues – and have developed the Slip Potential Model that allow owners of floors to predict through risk assessment where slips may happen.

The CIRIA guide is based upon this research and in particular the Slip Potential Model. It is designed to provide designers, procurers and managers of floors with good practice guidance that, if followed, should significantly reduce the number of accidents on existing and new floors.

Research is continuing. Work has started looking at human behaviour factors and also why people trip. This research will be used to update and extend the book, so that it becomes the authoritative guide to slips and trips.

Joyce Edmond-Smith
HSE Commissioner, with responsibility for slips, trips and falls

Acknowledgements

Research contractor   The research work and preparation of the guide were carried out by Ove Arup & Partners under contract to CIRIA.

Authors

John Carpenter

Currently a self-employed consultant, John was previously director of health and safety for Symonds Group (now part of Capita). He was co-author with Arup for CIRIA's *Safe access for maintenance and repair*. John's research work for HSE includes studies into undergraduate education, the provision of health risk guidance for designers, and competency and resource issues.

Deborah Lazarus

A structural engineer in Arup Research and Development, Deborah specialises in masonry, refurbishment work and general construction pathology. She is the joint author of CIRIA C556 *Managing project change: a best practice guide*, C579 *Retention of masonry façades – best practice guide* and its accompanying *Site handbook*, C589.

Clare Perkins

As a materials scientist in Arup Materials Consulting, Clare provides materials advice both to Arup worldwide and to external clients at all stages of the design and construction process. She specialises in timber, passive fire protection, waterproofing, plastics and rubber and architectural fabrics and has undertaken projects on the selection of stone and terrazzo flooring materials.

CIRIA manager   CIRIA's research manager was **Alan Gilbertson**.

Steering group   The study was guided by a steering group of experts representing parties involved in, or having an interest in, the prevention of slips on walking surfaces. CIRIA and Arup express their thanks to all the members of the steering group for their helpful advice and comments, and to the individuals and organisations who contributed information.

Chair   Michael Woods   Rail Safety and Standards Board

Members

Kevan Brassington	NBS
John Caves	Transport for London (London Underground)
David Ellis	Benoy (for British Council of Shopping Centres)
Les Fothergill	Office of the Deputy Prime Minister
Alan Gilbertson	CIRIA (secretary)
Susannah Gonzalez	Royal Institute of British Architects
Bob Keenan	Sheppard Robson
Jon Lawrence	QBE Insurance (for Association of British Insurers)
Paul Lemon	Health and Safety Laboratory
Marianne Loo-Morrey	Health and Safety Laboratory
Cindy Marshall	Tube Lines Limited
Jonathan Millman	NHS Estates
Colin Palmer	Henderson (for British Council of Shopping Centres)
Andrew Pitchford	CIRIA
Erica Ricks	NHS Estates
Richard Rogerson	Sandberg

Mike Roys	BRE
Tracy Savill	TRRL (for the Department for Transport)
Keith Snook	Royal Institute of British Architects
Geoff Street	BAA
Steve Sumner	Local Government Employers
Stephen Taylor	Health and Safety Executive
Stephen Thorpe	Health and Safety Laboratory

Corresponding members

Peter Jennings	ACO Technologies
Roger Barnard	Sport England
Jim Froggart	Football Licensing Authority

Workshop delegates

The CIRIA research project team also acknowledges all the following industry delegates, in addition to members of the steering group, at the industry workshop held at the Health and Safety Laboratory on 26 April 2005, who readily contributed their time and effort in answering the many questions put to them for discussion.

Dr Malcolm Bailey	Radlett Consultants
Julian Barnes	CERAM
Gary Bateman	Bonar Floors
Stuart Bell	Marshalls/BDA
David Browne	BAA
Mike Cox	J Sainsbury plc
Grant Currie	Buchanan Galleries, Glasgow
Peter Dolby	Isocrete
Peter Fereday	Pilkington's Tiles Group
Steve Ferry	SATRA
Clare Field	Health and Safety Executive
Malcolm Gilmore	Rossmore Group
Harry Gorton	Thortex
Chris Hughes	Rapra Technology
Alan Hunt	Kirklees Metropolitan Council
Peter Jennings	ACO Technologies/FACTA
Eddie Lloyd	Global Solutions Ltd
Alan McNeill	Kirklees Metropolitan Council
Pratibha Patel	Metronet
Vikki Pugh	Whitbread
Daniel Reynolds	SATRA
David Richardson	BRE/Stone Federation
Bill Robertson	Rossmore Group
Malcolm Stewart	Metal Flooring Manufacturers' Association
John Stinton	British Cleaning Council
Bill Tansley	Johnson Diversey
Carl Ward	Thortex
Pat Wherton	British Institute of Cleaning Services
Alan Williamson	Redman-Fisher
John Worth	Health and Safety Executive

Project funders

The project was funded by the Health and Safety Executive through CIRIA. Additional support was provided by:

Rail Safety and Standards Board

London Underground Limited

Estates and Facilities Division, Department of Health

Tube Lines Limited

Metronet

British Council of Shopping Centres.

Health and Safety Executive

This research project has been guided and facilitated by HSE. It would not have taken place without their representative's commitment and enthusiasm.

Stephen Taylor is a principal specialist inspector in HSE's Construction Division Technology Unit and is involved in all aspects of guidance, advice, scientific research and support for inspectors and the general public involving the safe use of buildings.

Health and Safety Laboratory

HSL has had a considerable input into this guide, primarily in terms of supplying and interpreting research material and making photographic material available. The contribution of those named below is acknowledged.

Dr **Marianne Loo-Morrey** is a higher scientist (materials scientist) in HSL's Pedestrian Safety Section. She is the principal investigator and project leader of research projects into aspects of the pedestrian tripping problem and of projects designed to gain a better understanding of the slip characteristics of a wider range of flooring materials.

Dr **Paul Lemon** is a senior scientist (physicist) within the Pedestrian Safety Section. He co-ordinates HSL's research efforts into the area of pedestrian safety and serves as the Laboratory's expert witness at legal proceedings. He previously led HSL's specialist slips and trips forensic investigation team.

Dr **Steve Thorpe** is a principal scientist and head of the Pedestrian Safety Section in HSL's Human Factors Group. He takes an active role in all the practical and management aspects of the diverse range of the team's work.

Note

Separate systems of building regulation apply in England and Wales, Scotland, and Northern Ireland. Similar intent exists under all three systems, though there are variations in both scope and method of application. For simplicity, this guides cites references to the system within England and Wales. Further information on building regulations in Scotland and Northern Ireland can be found online at, respectively:

<www.sbsa.gov.uk>

<www.dfpni.gov.uk/buildingregulations/technical.htm>

> ### Slips and trips training CD
>
> **Attached to the inside back cover of this book you will find a CD-ROM containing a Powerpoint presentation and training material on managing slips and trips.**

Contents

FIGURES

TABLES

Executive summary

Safer surfaces to walk on – reducing the risk of slipping provides comprehensive information and guidance on ways to reduce the number of slip incidents that occur on all types of interior and exterior walking surfaces. Slip incidents are responsible for 50 per cent of all reported accidents to members of the public. They result in a loss to employers and a burden on the health service amounting to hundreds of millions of pounds annually, and there is an enormous additional cost in terms of human suffering.

Production of the guide has been substantially funded by the Health and Safety Executive, aided by contributions from bodies responsible for the design, maintenance and operation of transport systems, shopping centres and hospitals – all organisations where the costs of slip incidents are high. It draws together research, much of it from the UK's Health and Safety Laboratory, to explain clearly how slips occur and the measures that should be adopted to prevent them. It is only quite recently that the detailed mechanisms of slipping have been understood, so this is the first time that such extensive guidance has been available.

The guide sets out the statutory obligations of those who specify and manage walking surfaces – requirements that may not be fully understood by those concerned. It is hoped that the guide will help these individuals and organisations meet their obligations.

The two preferred methods for measuring the properties of walking surfaces are described in some detail. Guidance is given on the range of values required for the different parameters to ensure that the walking surface has an adequate slip resistance under particular conditions. Case studies used throughout the guide illustrate practical applications of these measurements and of the management issues associated with floor maintenance. They also describe some forensic investigations where serious slip incidents have occurred.

A holistic approach to the selection of walking surfaces is required. The guide describes the Slip Potential Model, the so-called "gold standard" for selection, and also the Slips Assessment Tool developed by the HSE and promoted for initial design and comparative assessment. Flow charts are provided for the design processes involved in specifying new and refurbished walking surfaces. Examples of the application of these processes are also provided. The distinction is made between areas where control over factors such as use and footwear can be exercised, for example in factories and commercial food preparation premises, and those used extensively by the public. Areas that are used by both employees and the public are also considered; these include retail outlets and, importantly, healthcare facilities.

The guide illustrates that simple, cost-effective measures can reduce the incidence of slips. Failure to provide an appropriate surface can have serious financial implications. There is a need to bring together designers and facilities managers at an early stage of the design process. Consultation with the cleaning manager should also be undertaken. In specifying a walking surface the process of risk management should be followed.

Data is included on the slip resistance of a wide variety of flooring **materials** in both wet and dry conditions. Requirements for minimising the risk of slipping on specific building **elements** deal in some detail with stairs.

The impact of a wide variety of **contaminants**, both wet and dry, on the slip resistance of various walking surfaces is examined, including that of water. Although water might not commonly be regarded as a contaminant, it is probably the one most often encountered. Even a very small amount of water can dramatically reduce the slip resistance of a surface. Appreciation of the need to remove even small spills is emphasised, together with the need for the appropriate management strategy to ensure that this is implemented.

Some industries are known to have a particularly high incidence of slipping. The guidance draws on material on the HSE website addressed specifically towards the food processing industry, the food and drink industry and kitchens and food service. Guidance is also given for the education sector and the health services.

Managerial commitment to safety is identified as a pivotal influence on safety culture. Stemming from this, there is also discussion of the need for training to overcome lack of awareness on the causes of slips (and trips) and on the need to follow protocols on reporting incidents, cleaning spillages and using safety equipment. The guide recommends that **cleaning** should become an integral part of the work environment.

Much of the material used in the production of this guidance covers **slips** and **trips**, and also, in many cases, falls. Trips, primarily their causes and avoidance, are covered briefly in an appendix. It is intended that more detailed guidance will be produced on this topic before long.

Key issues

1 The scope of the guide

Slips cause untold pain and suffering every year, and are responsible for an unknown number of deaths arising from the complications that may follow. This guide is designed to provide comprehensive information and advice which, if followed, should greatly reduce the number of such accidents.

2 Target readership of the guide

The guide is intended for use by all those responsible for or involved in:

- specifying new floor materials for new and refurbished premises
- assessing the suitability of existing floor materials for changes in use
- managing the use, environment or maintenance (including the cleaning regime) of flooring surfaces
- investigating slip incidents on flooring surfaces.

3 Statistics (see also Appendix 3)

Slips and trips are the most common cause of injuries at work:

- 95 per cent of major slips result in broken bones
- they are estimated to cost the UK economy more than £1 billion a year in terms of staff replacements, claims and lost business; the human cost is incalculable
- *one slip or trip accident is estimated to happen every three minutes.*

4 Legal requirements (see also Appendix 2)

The law requires that walking surfaces be safe, that effective means of drainage be provided where necessary, and that contamination be avoided as far as reasonably practicable. The main legal requirements are set out in the Workplace (Health, Safety and Welfare) Regulations 1992; other legal requirements, such as the Building Regulations and health and safety legislation, will apply.

5 A framework for the specifier (see also Chapter 2)

The Slip Potential Model provides a framework for the specifier to give due consideration to all the factors that may influence the selection or refurbishment of a safe walking surface. Specifiers have to assess the risk of slipping, taking careful account of:

- possible *contamination* conditions or events
- the *cleaning* regime to be put in place
- the *floor surface material*
- *environmental factors* such as glare, colour, changes in surface condition or slope
- *human factors* including distraction, crowding, disability and encumbrance
- types of *footwear* that may be worn (in some areas the type of footwear may be within the control of some or all of the users; elsewhere, people may have no choice in the footwear they can use).

6   Risk management (see also Chapter 4)

The specifier has to follow the hierarchy of risk management when considering the factors influencing selection of a walking surface

7   The Slips Assessment Tool (see also Chapter 2)

The HSE Slips Assessment Tool may be used to for making approximate comparisons between different solutions in the preliminary stages of selection. It may also be used both to determine in general terms the slip potential of an existing floor and to monitor the wear of a surface in use.

8   Testing flooring materials (see also Chapter 3)

Testing a flooring material as it leaves the production line does not necessarily provide the necessary information for performance in us, because:

- the slip resistance value of the installed floor may differ from that of the ex-factory product
- surface roughness measurements of products in the factory may differ from those taken *in situ*, even immediately after laying
- some flooring materials will require surface treatment to provide good performance
- floors will wear with use, often becoming smoother over time
- contamination of surfaces – perhaps with even small amounts of water – *significantly* reduces the slip performance characteristics of most floors and increases the risk of slipping.

9   Alternative test methods (see also Chapter 3 and Appendix 5)

- There are many methods for testing the performance characteristics of flooring materials, some of which are more reliable than others, depending on the circumstances
- each material has different characteristics under different conditions, and the testing of floors may be considered a science in itself
- the interpretation of the test results may not be readily apparent: similar numbers from different tests will not have the same interpretation.

This guide explains the various tests that may be used and their relative merits but suggests the pendulum test and the surface roughness meter as the preferred methods.

10   Facilities maintenance (see also Chapters 5 and 6)

Those responsible for the use and maintenance of floors need to take account of the factors affecting slip resistance, and in particular to recognise that:

- cleaning should be integral to the work environment
- manufacturers' instructions for cleaning specific flooring products should be followed, together with any requirements contained within the maintenance plan
- surfaces should be thoroughly dried after cleaning and before subsequent use
- contamination should be removed promptly and thoroughly to leave a clean, dry surface
- testing for change of roughness over time should be carried out, with further pendulum testing considered when significant changes have been noted. Areas that have become too smooth will be need to be renewed or replaced.

11   Strategic risk management policy

Managerial commitment to safety is a pivotal influence on safety culture:

- a risk management strategy policy should be implemented by those responsible for the premises
- appropriate (and adequate) training is required
- near-miss reporting should be incorporated in the policy
- slip investigations should aim to identify the root causes.

12   Human factors (see also Chapter 9)

The contribution of human factors to the occurrence and prevention of slips needs to be recognised:

- slip incidence can be reduced by modification of behaviour, which may be achieved with appropriate training and/or warning of potential hazards
- distractions cause people to overlook obvious hazards
- impaired mobility when pushing, pulling or carrying contributes to a large number of slips, trips and falls
- age is a significant factor in many slip accidents
- physical frailty is a factor. Vulnerable groups include the disabled, children, and young people in their first employment.

13   Flooring surface properties (see also Chapter 10)

Profiled surfaces, including ceramic tiles, pressed metal sheet or rubber sheet, are perceived as being slip-resistant, but this is not necessarily the case. Unless there is interlock between the shoe and the profiled surface, it is the surface micro-roughness of the highest surface of the profile that determines the slip resistance (see also Chapter 5). Pendulum testing of profiled surfaces is possible, but the data generated may need careful interpretation: expert advice should be taken. The HSL ramp method is an alternative test method for profiled surfaces.

14   Surface compatibility (see also Chapter 11)

Elements inserted within a floor can constitute slip hazards, where adjacent surfaces have very different slip resistance. Examples include a metal cover for a drainage channel set in paving, or a smooth metal or plastic ventilation cover in a carpeted floor.

15   Building entrances (see also Chapter 11)

Building entrances require special attention to reduce the likelihood of slip incidents:

- the location and design of the entrance should aim to reduce the extent to which water and dirt are carried into the building, whether blown by wind or transported by users. Canopies and lobbies that provide shelter from prevailing weather and act as an intermediate zone between outdoors and indoors can be effective
- the provision of appropriately located, sized and maintained matting is also of great importance.

16 Stairs (see also Chapter 11)

Serious slips and falls often occur on stairs:

- factors in stair design that are particularly relevant to slips and trips are the size of the tread, the shape of the nosings, the inclusion of proprietary nosings and the slip resistance of the tread or nosing material where first contact with the foot is expected in descent

- a small tread size greatly *increases* the risk of slipping caused by overstep in descent. Conversely, larger treads significantly *reduce* the likelihood of slipping in descent

- the risk of slipping also increases if the tread and nosing are finished in a smooth material, if the steps are wet, or if the edge of the step is rounded (reducing the effective tread size)

- poor lighting and a lack of usable handrails may also increase this risk.

17 Footwear (see also Chapter 7)

Footwear, and the extent to which this can be controlled, is the component of the Slip Potential Model to be considered in determining the required slip resistance of the floor surface: the properties of the shoe sole are highly relevant to reducing pedestrian slipping:

- the wear rate and to a degree, cleanability of the sole influences the surface roughness levels throughout the life of a shoe sole and is vital in maintaining satisfactory slip resistance in contaminated conditions.

It is important that footwear fits correctly, as slipping is more likely if the wearer's foot moves within the shoe.

Glossary

contamination	Any substance (wet or dry) on the walking surface that may reduce the slip resistance
going	The horizontal distance between the **nosing** of a tread and the nosing of the tread or landing immediately above it
HSL ramp test	An HSL laboratory test derived from a DIN standard
major injury	As defined in Schedule 1 of the Reporting of Injuries, Diseases and Dangerous Occurrences Regulations 1995 (RIDDOR)
macro-rough	Surfaces with embedded but protruding aggregate or similar material
micro-rough (or rough)	The irregularities in a walking surface, often invisible to the naked eye, and measured at the micron (μm) scale. See also **rough**
nosing	The front edge of the tread (upper surface) of a step
pedestrian	For the purposes of this guide, an individual traversing the walking surfaces defined in Figure 1.2
pendulum test	The preferred means of establishing slip resistance
profiled surface	A surface with a designed geometric vertical profile
pendulum test value (PTV)	A term used in BS 7976 *Pendulum testers*. It means the same as **slip resistance value**, the term used in this guide
rough (or micro-rough)	Surfaces with a surface roughness (Rz) typically between 10 μm and 100 μm as measured by a surface roughness meter
roughness (Rz)	The measurement of average peak-to-trough height as measured by a micro-roughness meter. Equivalent to **micro-roughness**
sleds	Instruments designed to measure slipperiness by rolling across the walking surface. Generally considered by HSL and HSE to be inferior to the **pendulum test**
Slips Assessment Tool (SAT)	A PC-based package recently developed by HSE and HSL that allows non-experts to assess the slip risk potential presented by level pedestrian walkway surfaces
Slip Potential Model	A framework derived by HSE/HSL to encourage a holistic approach to the assessment of slip potential
slip resistance value (SRV)	The result from the pendulum test where the test is performed using the Four-S rubber slider. Equivalent to the **pendulum test value**
specifier	The person who dictates the choice of walking surface material and/or the other controllable influences. This may be a client, designer, architect, facilities manager, quantity surveyor or a contractor

Abbreviations

ACOP	approved code of practice
ADB	Approved Document B (The Building Regulations 2000)
ADK	Approved Document K (The Building Regulations 2000)
ADM	Approved Document M (The Building Regulations 2000)
AFARP	as far as reasonably practicable
CDM	The Construction (Design and Management) Regulations 1994
CoF	coefficient of friction
COSHH	The Control of Substances Hazardous to Health Regulations 2002
cP	centipoise
DDA	Disability Discrimination Act 1995
DIN	Deutsches Institut für Normung (German Institute for Standardisation)
HSC	Health and Safety Commission
HSE	Health and Safety Executive
HSL	Health and Safety Laboratory
LA	local authority
MHSWA	The Management of Health and Safety at Work Regulations 1999
NFSI	[US] National Floor Safety Institute
PTV	pendulum test value
PUWER	The Provision and Use of Work Equipment Regulations 1998
RIDDOR	The Reporting of Injuries, Diseases and Dangerous Occurrences Regulations 1995
RSSB	Rail Safety and Standards Board
R_{pm}	average peak height (also known as peak roughness parameter)
R_{tm}	total roughness parameter
R_{vm}	average valley depth (also known as valley roughness parameter)
Rz	surface roughness
SAT	Slips Assessment Tool
SRA	Strategic Rail Authority
SRV	slip resistance value
STF	slips, trips and falls
UKSRG	United Kingdom Slip Resistance Group

1 Introduction

> **KEY POINTS**
>
> ◆ *Slips are an important and dominant element in the overall pattern of accident statistics.*
>
> ◆ *A holistic approach should be adopted to understanding and analysing slips.*
>
> ◆ *There are key drivers encouraging the improvement of the slip resistance of walking surfaces.*
>
> ◆ *The guide contains wide-ranging background data to assist specifiers and others.*
>
> ◆ *The readership of this guide should include all those with an interest in and responsibility for walking surfaces.*
>
>
>
> Courtesy HSL

1.1 GENERAL

This guide emanates from a recognition that injuries to pedestrians as a consequence of slipping on walking surfaces is a major cause of personal suffering, disruption, and cost to individuals, organisations and ultimately to the national economy. It is the most common cause of injury in the workplace. For employers, and owners and managers of walking surfaces, the likelihood of slip accidents should be a major concern for the reasons set out in this guide.

The consideration of slips is usually as part of the grouping "slips, trips and falls". This guide however concentrates on "slips" with the aim of collating the key research into this area, highlighting current thinking as to causality, and providing advice to clients, owners, designers, managers and others with a responsibility for the provision, specification and maintenance of safe flooring surfaces.

Slips would probably not be recognised as one of the major safety concerns in many businesses and organisations, and many people would perhaps consider them rather mundane. However anecdotal evidence suggests that, in 2003, there were almost 900 000 NHS bed-nights arising from slips, compared with 82 000 from car accidents. Statistically, it is one of the most common risks confronting us all (Beaumont *et al*, 2004). The simple message is:

- slips can have serious consequences

- simple, cost-effective measures can reduce these accidents.

This serious message is not assisted by a supposedly comic side to slipping; it is after all the backbone of many comedy sketches, particularly those directed at children. It is these children who are the managers of tomorrow.

People with an impairment, through age, disability or perhaps by virtue of what they are carrying or manoeuvring, are particularly prone to this form of accident. As the demographic shift continues over the next few decades, age-related impairment is likely to generate a significant increase in slip-related injury.

The elderly are particularly susceptible to slips as they are far more sensitive to the various contributory factors, illustrated in Figure 2.1. They are also more likely to suffer injury than a younger person; either a direct injury, such as a broken hip, or a consequential medical complication, which in some cases may ultimately result in death. Visual acuity decreases with age (Newton, 1997). Stairways, steps and uneven surfaces can therefore be a problem for people with decreased depth perception, ie a diminished ability to judge distances (Beaumont *et al*, 2004).

This guide aims to dispel some of the biggest barriers in respect of slip accidents, ie:

- not taking the risks seriously

- not understanding the causes of slipping

- thinking that slips are inevitable

- poor application of risk assessment and management controls.

1.2 BACKGROUND STATISTICS*

Slips and trips are the most common cause of major injuries[†] at work. They occur in almost all workplaces; 95 per cent of major slips result in broken bones and they can also be the initial causes for a range of consequential accident types such as falls from height[‡] or subsequent ill health in the elderly.

It is a sobering thought that, on average, workplace slips and trips account for:

- 33 per cent of all reported major injuries

- 20 per cent of over-three-day injuries to employees

- two known fatalities per year (as a consequence of consequential effects, eg falling from height)

- 50 per cent of all reported accidents to members of the public

- incalculable human cost (HSE website, <www.hse.gov.uk/slips>).

---

*   Where separate statistics are not available, they are quoted for both slips and trips.

†   Defined in Schedule 1 of the Reporting of Injuries, Diseases and Dangerous Occurrences Regulations 1995.

‡   HSE estimates that 30 per cent of falls from height are initiated by a slip or trip.

Although only two fatalities per year are reported in the data above, it is suspected that slips are often the root cause of fatalities arising from the consequences of a slip, particularly in the elderly. Hence the true fatality rate is almost certainly higher. In addition these figures do not include non-workplace injuries and fatalities (particularly in the home), which account for a substantial number of slip accidents.

The breakdown of statistics does not allow a detailed analysis of slips alone. However since additional data became available in 2002/03 it is estimated that 19 per cent of all slips, trips and falls were attributable to slips on surfaces that were "wet or covered in substance" (HSC, 2004). Because of factors such as the under-reporting of accidents in many industry sectors, and the tendency of the sufferer's clothing to absorb contamination after slips (thus hiding or disturbing the evidence), it is believed that the real figure could be higher. A study carried out to identify human factors associated with slip and trip accidents (Peebles *et al*, 2004) indicated that slips accounted for 46 per cent of all the reported incidents.

Some sectors have a particularly poor record in this area, specifically the food, drink and tobacco industries (HSE, 1996a). The hospitality sector is also poor and suffers from a very low formal accident reporting level. Recent research (HSL, 2004) in respect of trips has concluded that "the service industry should be the focus of attention in efforts to reduce the number of trip accidents".* The same could be said of slips.

The concern is not just a human one, important though that is. The costs associated with slips and trips are significant and amount to:

- £512 million per year for employers
- £133 million per year for the National Health Service.

A reduction in the incidence of slips therefore has the direct benefit of improving efficiency, profits and the nation's wealth.

Statistics are covered in more detail in Appendix 3.

1.3   SPECIFYING A WALKING SURFACE: A HOLISTIC APPROACH

Historically, the need for walking surfaces to provide adequate resistance to slipping has not always received sufficient attention, with specifiers often prioritising cost and appearance. Those specifiers who have given the issue some thought, have in some instances made conditions worse because they have had an incorrect appreciation of the mechanism of slipping. To be fair, a realistic understanding of the many strands of contributory factors necessary to avoid or minimise the occurrence of slipping has only recently been developed, and is still in its formative stages of verification, although results are very encouraging. A further complication has been the lack of coherence and reliability in the manner in which manufacturers' data has been presented.

Some commentators (Thorpe and Lemon, 2000; Rossmore Group, 2003) suggest that taking a holistic approach in tackling these issues enables the causes of slips to be determined. This will allow mitigation plans to be developed that tackle root causes rather than assuming that slips are inevitable. These various factors interact, so it is not possible to give a general rule – each case needs to be considered on its own merits. This is the "goal setting" approach required by health and safety legislation.

---

*   The *Review of RIDDOR trip accident statistics 1991–2001* (HSL, 2004) is concerned with trips. There is no equivalent study for slips, but the sentiments are likely to be the same.

The holistic approach considers the causes as being made up of the dynamic relationships between the environment, the processes or tasks that are carried out within that environment, the technologies present and in use, and finally the users or people that use them (Rossmore Group, 2003). The last element is a complex part of the system since they must be considered as both physical and psychological elements of the whole. This process is discussed in Chapters 2 and 4 in particular.

The guide has been generated from a large quantity of research data derived by HSL, and expert advice from HSL, HSE, the steering group and others. This data has been open to expert scrutiny and has been presented at industry workshops. Nonetheless, it is intended that the guide will be kept under review, so feedback is welcomed.

The "Swiss cheese" model (Reason, 1997) is a good illustration of how accidents can occur, and indeed can be prevented. The theory postulates that each "slice" is a defence against an accident; the holes represent local shortcomings in these defences. When a number of holes manage to align however, it indicates a situation where all the defences have been breached and an accident may occur. These "slices" are represented by the many factors which work to prevent slips. These are explained in Chapter 2.

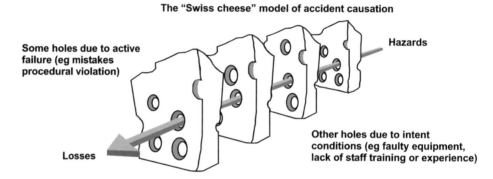

The "Swiss cheese" model of accident causation

Some holes due to active failure (eg mistakes procedural violation)

Hazards

Other holes due to intent conditions (eg faulty equipment, lack of staff training or experience)

Losses

Successive layers of defences, barriers and safeguards

Figure 1.1   *Swiss cheese model of accident causation (Reason, 1997)*

It has not been appreciated by many specifiers or managers that there is an obligation on them to ensure that floor surfaces are suitable for their purpose – ie they must not be slippery.* This has to be achieved in part through the hierarchy of risk management.† This has two implications:

- there is a sequence that must be followed in order to satisfy these regulations

- as new technology or practice becomes available, and societal expectations rise, these must be taken into account in deriving a solution.

The second bullet point means that what might have been an acceptable solution in the past, may not be acceptable in the future on account of developments in materials, cleaning technology, or some other advance. All those involved therefore need to keep abreast of contemporary thinking.

The key to success is to get it right first time.

---

*   The Workplace (Health, Safety and Welfare) Regulations 1992 infer that this is an absolute obligation; case law has identified some latitude, however (see Section 9.9.1).

†   As set out in the Management of Health and Safety at Work Regulations 1999.

DRIVERS

Some of the drivers that have created a need for improved understanding are listed below.

Health and safety-related legislation

Health and safety legislation makes it clear that those specifying walking surfaces, and those who subsequently manage them, have a statutory obligation to ensure that they are safe. Quite what constitutes "safe" has been a matter of debate, but this guide is intended to assist in this respect. Appendix 2 outlines the relevant acts and regulations that apply to the safety of walking surfaces.

The Disability Discrimination Act 1995 (DDA)

This ground-breaking Act requires those responsible for premises accessible to the public to make reasonable adjustments to prevent discrimination against those with disabilities. The two key impairments relevant to the safe use of walking surfaces are sight and walking gait. This Act indirectly affects both specifiers and manufacturers of new flooring and directly affects those who manage existing surfaces. The DDA exempts by regulation any feature of a building that complied with the relevant objectives, design considerations and provisions of the edition of Building Regulations Approved Document M that applied at the time of construction and continues substantially to do so, provided no more than 10 years have elapsed since the construction of that feature.

Civil action

Although the assertion that we live within a "compensation culture" is not universally accepted*, many organisations with responsibilities for walking surfaces pay out thousands of pounds each year following claims arising from injury sustained from poor surfacing (albeit not all of them arising from slips). There are various reasons why a responsible duty-holder should be concerned, and one of them is surely the cost of these actions. Adopting a whole-life approach often demonstrates the sense of choosing the correct surface at the start.

Costs

Apart from the potential costs arising from a civil action, incorrect choice or maintenance of a walking surface can have significant cost implications for owners and operators. These arise both as direct costs, such as the expense of replacing or re-treating the surface, and also as indirect costs arising from disruption, damage to the organisation's image, loss of trade and inefficiencies in working etc. The latter group can often be significant.

---

* Keynote speech by Lord Falconer at HSC seminar on risk and compensation, 22 March 2005.

Higher standards and best practice

Many of our common public areas are no longer considered to be low-priority streetscapes: shopping malls, station concourses, airport terminals and hard landscape areas have high-quality walking surfaces. Architects and others are constantly striving for improvements both in quality and cost. These materials often feature polished surfaces. While this general advance is to be welcomed, it is necessary for all those concerned (manufacturers, specifiers and maintainers) to realise that they must also be safe. This guide emphasises that this is achieved through the management of a wide range of influencing factors, which demands the holistic approach mentioned above.

This improvement is matched by a corresponding increase in industry standards and guidance, particularly from clients and owners with large interests such as Network Rail.

HSE initiatives and accident statistics

HSE is very much aware of the importance of slips (and trips and falls) as components of ill-health and accidents generally and it has a priority programme directed at the topic. The statistics given in Section 1.2 and Appendix 3 support this. HSE's website, <www.hse.gov.uk>, contains data and background information on the subject, and leaflets and guidance documents are available from HSE Books. HSE has developed the Slips Assessment Tool (SAT), which is described in Chapter 2.

HSE set overall targets for accident reduction in 2000 (HSC, 2000), which included reducing the rate of fatal and major injuries to workers by 5 per cent by 2005, and by 10 per cent by 2009/10. This target does not include those not at work such as the elderly. Although an accurate analysis of accidents over the period since 2000 is not possible owing to changes in the way accidents are reported and analysed, it appears that the accident rates are being maintained or are increasing slightly (HSE, 1996a). Given the prevalence of slips (trips and falls) within the overall accident rate, HSE has a continuing concern over the level of slip accidents.

Industry initiatives

The UK Slip Resistance Group (UKSRG) is made up of representatives from the flooring and associated industries, including floor manufacturers, representatives of the HSE, test houses, end-users, instrument manufacturers and forensic engineers. In October 2005 UKSRG issued the third edition of its guidelines for the testing and investigation of walking surfaces (UKSRG, 2005).

WALKING SURFACES CONSIDERED IN THIS GUIDANCE

This guide considers a wide range of walking surface situations, as illustrated in Figure 1.2, which also indicate some exclusions that should be noted.

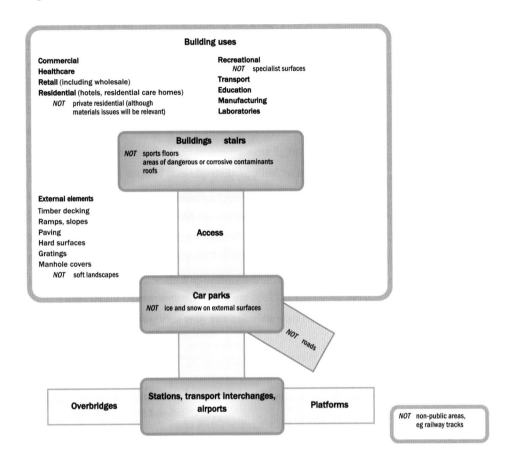

Figure 1.2   *Surfaces considered in this guide*

1.6   BACKGROUND DATA

Although the research team has considered a wide range of UK and overseas research material, the primary source and emphasis for this guide has been the research carried out by the Health and Safety Laboratory and the Health and Safety Executive.

The period of research in this area of investigation is relatively recent and it is only over the latter part of this time-span that an understanding of the mechanisms of slip accidents has started to become sufficiently clear to allow theories to be postulated with confidence. The Slip Potential Model and the Slips Assessment Tool (both described in Chapter 2) utilise this understanding, but with an awareness that more research is required in many respects. As this knowledge becomes available the Slip Potential Model and the Slips Assessment Tool will need to be adapted accordingly.

This guide is designed to give an authoritative background to slips and to help those who:

- specify new floor materials for new and refurbished premises
- assess the suitability of existing floor materials for changes in use
- manage the use, environment or maintenance (including cleaning) of flooring surfaces
- investigate slip incidents on flooring surfaces
- need to identify the ways in which various factors interact and thereby potentially affect the safety of the surface.

The guidance will be of interest therefore to:

Clients/owners	Promoters of projects, owners of facilities, those responsible for facilities
Designers/specifiers	Design consultants, specialist contractors, manufacturers
*Planning supervisors**	To provide a base from which actions of others may be judged and questioned
Maintenance contractors	Facility managers, cleaning companies
Forensic architects and engineers	Those who need to review existing floors to assess suitability or investigate accidents
Others	Tenants, landlords

Chapter 2 provides an outline of the physiology of walking and describing the two main techniques used to derive a safe surface. The importance of manufacturers' data is also discussed. **Chapter 3** describes the testing methodologies favoured by HSE/HSL and their application to various generic surfaces with wet or dry contamination.

Chapter 4 outlines the obligations of those with an involvement in the specification or management of walking surfaces. It explains how the techniques described in Chapter 2 may be used to practical effect, drawing on data covered in the other chapters.

Chapters 5–10 review the various influences illustrated in Figure 2.1 and that contribute to the Slip Potential Model, namely contamination (**5**), cleaning (**6**), footwear (**7**), environment (**8**), human factors (**9**) and surface characteristics (**10**).

Chapter 11 describes building elements relating to walking surfaces. The guide concludes with **Chapter 12**, which provides a number of case studies. Supporting data is contained in **Appendices 1–6** including a brief commentary on trips in **Appendix 4** and a review of other surface test methods in **Appendix 5**.

---

*   Due to be replaced by the "co-ordinator" in the 2006 draft revision of the CDM Regulations.

2 Setting the scene

> **KEY POINTS**
>
> ◆ *The physiology of walking is relevant to an understanding of slip accidents.*
>
> ◆ *The Slip Potential Model (SPM) and the Slips Assessment Tool (SAT) are the two preferred methods of analysis.*
>
> ◆ *Manufacturers' data is useful and improving, but should be employed with caution.*

Courtesy Deborah Lazarus

2.1 THE PHYSIOLOGY OF WALKING

The physiology of walking is complex and has been investigated by a number of researchers. This chapter gives an outline of the key factors. A more detailed description is given by the Rail Standards Safety Board (Beaumont *et al*, 2004).

Stability

Five factors define and determine stability during walking (Hyde *et al*, 2002):

- body weight
- height of the centre of gravity above the support base (the area spread by the feet and any other support, such as a walking stick)
- size of the support base
- relative motion between the centre of gravity and the support base
- whether the line of action of the centre of gravity (the gravity line) is within or outside the support base.

Slipping is defined as a fall caused by sliding when there is a sudden loss of all or part of the support base such that the gravity line moves outside the remaining support base.

Gait analysis

Gait analysis is the systematic study of human walking. Each step can be considered to be a gait cycle with seven phases, starting where initial contact is made by the heel of the right foot. An understanding of the components of this cycle assists in the interpretation of slip mechanics. Slipping is most likely to occur just after the heel makes initial contact with the ground or as the toe pushes off.

Gait analysis in the elderly

Several investigations have been undertaken on changes in gait with increasing age (Murray, 1967; Murray et al, 1969). Age is a significant factor in falls as a gradual decline in balance and speed of gait occurs with age. Combining age and sex factors has shown that elderly women fall more often than elderly men.

The prevention of slipping is therefore concerned with human physiology in addition to the physical aspects of the floor, its management and the immediate environment.

Gait is also discussed in Section 9.4 in connection with human factors.

2.2   SLIP POTENTIAL MODEL

Research over a number of years by HSE and HSL in particular (HSE, 2004a; Thorpe and Lemon, 2000) has shown that a combination of factors contribute to pedestrian slip accidents and that it is important to consider all of them when investigating, selecting or managing walking surfaces. In the light of this a Slip Potential Model has been developed in which the factors in Table 2.1 are assessed in a given situation.

Table 2.1   *Factors included within Slip Potential Model*

	Factor	Chapter in which discussed
1	Contamination	5
2	Cleaning	6
3	Footwear	7
4	Environment	8
5	Human (pedestrian) factors	9
6	Floor surface material	10

Figure 2.1 demonstrates how these influences may be separated into those that are generally controllable, and those that are not controllable, but are largely predictable.

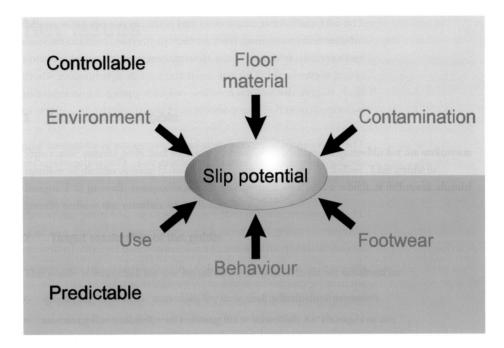

Figure 2.1 *Slip influences*

These factors are discussed in detail elsewhere in this guide and all play a part, in particular, in the selection or analysis of walking surfaces described in Chapter 4.

The essence of the Slip Potential Model approach is the recognition that a number of issues contribute to the potential for pedestrian slip accidents; it is not sufficient to consider one or more in isolation. It is by understanding the inter-relationship, and the relevance, of each component in a particular circumstance that an overall assessment of the slip potential may be made. This holistic strategy is considered to be the definitive method within the current state of understanding. The model does not provide an absolute answer, but assists with decision-making. As new research becomes available an increasing ability to quantify the mechanisms of slip, and to make choices accordingly, may be possible.

So far as the walking surface itself is concerned the key attributes are, first, its dynamic coefficient of friction (from which the slip resistance value (SRV) value is obtained) and, second, the surface roughness (Rz). The derivation and use of these indicators are described in detail in Chapter 3 and discussed throughout the guide.

The application of the model to the selection and management of walking surfaces is outlined in Chapter 4.

2.3 SLIPS ASSESSMENT TOOL (SAT)

HSE/HSL have also developed a software package to assist those who are not expert in this field to assess the slip potential presented by level pedestrian walking surfaces. It can be found on <www.hsesat.info> (HSE, 2004a). The SAT is based on the Slip Potential Model described above. It is not considered as rigorous as the use of the model itself, as:

- it contains predetermined value judgements
- does not utilise the dynamic coefficient of friction, which is the preferred means of characterising walking surfaces (utilising instead the roughness parameter).

The SAT package is easy to use and prompts the operator to gather relevant information concerning floor surface properties, contamination, cleaning regimes, footwear etc. When all of the information has been entered into the package, a slip risk rating is produced. This will assist the user in determining whether site conditions are likely to give rise to a high or low risk of slipping. This emphasises the need for a holistic approach to the assessment of slip resistance in a given location.

The data can be entered into a laptop computer (pre-loaded with the SAT software) on site for an immediate assessment of slip risk. This is the preferred method of operation. Alternatively, data can be recorded on site using a proforma and entered into a PC later.

The assessment can then, if desired, be repeated using alternative data such as a different cleaning regime or footwear type etc. This will produce a different (theoretical) slip risk rating. This is considered by HSE to be a powerful way of demonstrating the relative importance of various slip risk control measures and might usefully be employed in the early design stages.

HSE points out that the program "must not be used as the basis for floor surface specification or modification, but rather as a simple aid to identify generic levels of slip risk". It does not take account of sloping surfaces, the capabilities of the pedestrian, pushing or pulling action, and other factors.

The SAT is useful for monitoring changes over time, where it is the comparison of results that is important rather than absolute values. Overall it is a helpful guide, rather than providing a definitive approach to the prediction of slipperiness or the analysis of an accident. For the latter, the Slip Potential Model, using forensic evidence from the pendulum test, is recommended and it is considered essential for derivation of data for use in litigation proceeding.

It should be noted that the numerical ranges given on completion of the SAT analysis are not equivalent to the SRV values obtained utilising the pendulum test (Section 3.2).

The SAT reflects available research data and understanding of the slip mechanisms, presented in a simplified format within the software. As more experience and data is gained HSE intends to update the model and so improve its effectiveness.

2.4 MANUFACTURERS' DATA

Walking surface materials and footwear will often be accompanied by manufacturers' data in respect of resistance to slips. Most of this information relates to the products "as supplied", so it does not take account of:

- changes introduced during the construction or laying process, eg mortar splashes, grinding or polishing
- degradation, wear, maintenance and use

all of which can significantly affect the actual resistance to slip. Some manufacturers have begun to include in their literature data referring to the change in slip potential of the floor material after prolonged use. This is to be welcomed.

In addition, the friction parameter quoted should be viewed with caution, as the means by which it was derived can have a critical effect upon its validity.

Data can also be presented in a potentially misleading format: for instance, quoting "R9" slip resistance to DIN standards as being "good slip resistance" when in fact this result represents the lowest (most slippery) result achievable on this scale (see Appendix 5).

Testing of walking surfaces

KEY POINTS

- *There are a several assessment instruments available to measure surface properties.*

- *The preferred methods are the pendulum test, supplemented by the roughness meter. The HSL ramp test is recommended for specific test requirements.*

- *Research has shown that other test methods can give misleading results, but HSL is continuing to evaluate them.*

- *Walking surfaces should be generally described by their slip resistance (SRV) supplemented by the surface roughness (Rz).*

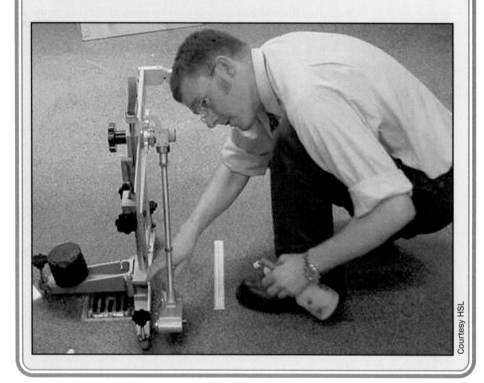

Courtesy HSL

3.1 ASSESSMENT INSTRUMENTS AND TECHNIQUES: WALKING SURFACE PROPERTIES

As is mentioned in Section 2.2, the key properties of a walking surface in respect of its capacity to resist the potential for pedestrian slips, are the dynamic coefficient of friction (CoF), usually presented as its equivalent slip resistance value (SRV), supplemented by the surface roughness (Rz). In most circumstances, and particularly where a slip accident has occurred, it is recommended by HSE that both these parameters (SRV and Rz) are used to give an accurate indication of floor surface slipperiness.

There are various methods by which these properties may be measured, but research by HSE/HSL and others over the years has convinced them that the three approaches described below:

- pendulum test (to measure dynamic coefficient of friction)
- surface roughness measurement

and, to measure footwear, specific surface slipperiness or bespoke combinations,

- HSL ramp test

are the optimum techniques. The first two tests are also recommended by the UK Slip Resistance Group (UKSRG, 2005).

For proposed walking surface materials these tests are normally applied in laboratory conditions; for existing walking surfaces the location of the test measurements will depend upon the nature of the study, but sufficient test areas should be chosen to give a representative picture of the floor surface. For example:

- at the site of an accident or area of concern
- in heavily trafficked or worn areas
- across the width of passageways or stairs to determine the wear pattern
- in areas with little use (eg behind or underneath furniture, vending machines. and the like) to provide a benchmark of "original parameters".

3.2 PENDULUM TEST

It is the dynamic coefficient of friction, rather than the static coefficient of friction, that is considered to be the most relevant in the assessment of walking surfaces as this reflects the action of the heel. The pendulum test replicates this action on a contaminated surface,* in particular its interaction with the hydrodynamic squeeze film (Section 5.4).

The measurement of the dynamic CoF is achieved using the pendulum test in accordance with BS 7976 Parts 1–3:2002 *Pendulum testers*, which covers the specification, method of operation and method of calibration of this equipment, together with the testing protocol outlined in the UKSRG guidelines, the third edition of which was recently issued (UKSRG, 2005).

This pendulum test is also known as the portable skid resistance tester, the British pendulum and the TRRL pendulum (HSE, 2004a).

The pendulum test was designed to simulate the action of a slipping foot, although it has also been used to measure the skid resistance of roads. The method is based on a swinging dummy heel that impacts and sweeps over a set area of flooring in a controlled manner.

The slipperiness of the flooring has a direct and measurable effect on the pendulum value determined. This is known as the slip resistance value (SRV), pendulum test value (PTV[†]) or British pendulum number. For most of the test range encountered there is a simple relationship between the coefficient of friction and the SRV[‡]. The more slippery the surface, the lower the SRV.

---

* The background to this test, and its adoption by HSE/HSL as the preferred method for the site measurement of the dynamic CoF, is reviewed in Bailey (2005).

† The term used in BS 7976-2.

‡ SRV is equivalent to 100 × the CoF.

Figure 3.1   *Pendulum test instrument (courtesy HSL)*

Notwithstanding the adoption of this test by HSE/HSL as the preferred model, it requires a competent operator both in its use and in the interpretation of the results. However, HSE/HSL currently believe (HSE, 2004a) that this is the only portable instrument that simulates the fluid dynamics of a foot slipping on a wet-contaminated floor. Although new instruments are being developed, bringing forward new ideas, neither SlipAlert nor the SATRA test generates suitable impact dynamics (these tests are discussed in Appendix 5). At present, the pendulum remains the preferred option.

Standard pendulum results are given in relation to the use of a standardised soling material on the pendulum "foot". The standard Four-S rubber (standard simulated shoe sole) was developed as a rubber of average slip resistance characteristics. For the assessment of barefoot areas, or unusually rough floors, the use of the TRRL rubber (a similar but softer, more malleable compound) may be advantageous. Slider choice need not be limited to either material, provided the chosen material is used consistently and it is recognised that comparisons with the results from standard rubbers will not be possible if alternatives are used.

HSL has investigated rubber compounds (Loo-Morrey and Hallas, 2003) used on the German GMG100 test apparatus* for possible applicability to the pendulum test. The results indicated that it did not reliably distinguish between:

- similar surfaces, because of the reduced range of slip resistance values measured when using the pendulum

- wet, dry or soapy floors (two floors were measured to have higher coefficient of friction when either wet or soapy, than when dry) when using the GMG100.

HSL concluded that GMG100 should not be adopted for use on the standard pendulum test slider, and thereby confirmed the unsuitability of most sled-type tests for assessing the slip resistance of floors.

---

*   Supplied by GTE-Industrieelektronik, <www.gte.de>.

Although not ideal, because of the practicalities of setting up the instrument on a slope, the pendulum has been found to give accurate results on inclined surfaces so long as the pendulum frame itself is kept horizontal. It can be used within a slope range of ±10°. However, it only assesses the material's slipperiness and not the extra risk of slipping as a result of the incline (HSL, 2002).

The pendulum is not normally considered suitable for use on stair treads and nosings because of the size of the test area required. In addition, the use of the pendulum on heavily profiled flooring materials requires operators with experience; the setting up of the instrument and the interpretation of the results demand particular care and skill. Consideration of alternative test methods, such as comparative roughness tests (for stairs) and ramp tests (for heavily profiled surfaces) should be considered in such cases (Lemon, Thorpe and Griffiths, 1999a); see also Section 3.5.

The pendulum test may be used on carpet surfaces, although slips on carpets are not generally an issue provided the carpet is of adequate quality, well fitted and maintained.

Table 3.1 *Pendulum results (SRV) in relation to slip potential: Four-S rubber (from Table 1, HSE, 2004a)*

Pendulum value (SRV) (Four-S rubber)	Potential for slip*
0–24	High
25–35	Moderate
36–64	Low

*   Denoted slip risk in original reference. The risk categories are quantified in Section 4.2.1

If the TRRL rubber is used (HSL, 2000), slip potential may be estimated as shown in Table 3.2.

Table 3.2 *Pendulum results (SRV*) in relation to slip potential: TRRL rubber*

Pendulum value (SRV*) (TRRL rubber)	Potential for slip**
0–19	High
20–39	Moderate
40–74	Low
75+	Extremely low

*   The term SRV is strictly only applied to pendulum values using Four-S rubber
**   Denoted slip risk in original research. The risk categories are quantified in Section 4.2.1

Standard technique (wet contamination)

UKSRG recommends that the investigation is carried out on an area of floor of approximately 500 mm × 500 mm using at least six test areas (150 mm × 100 mm) to accommodate wet and dry conditions, in three directions (see Figure 3.2). The test area should be used only once. If the flooring material has directional properties the testing methodology will reveal this.

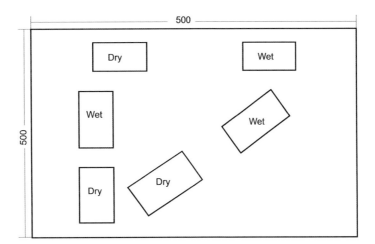

Figure 3.2   *Schematic of test area on a test site of approximately 500 mm × 500 mm (from UKSRG, 2005)*

Pendulum tests are carried out in a series of five swings, following three initial conditioning swings. The use of the conditioning swings allows a degree of interaction to occur (Lemon *et al*, 2001) between the test surface and the rubber pad of the pendulum and generally results in a marked improvement in the consistency of the test swings. The stated SRV value is the arithmetic mean of the five test swings.

Initial tests would normally be carried out on the floor surfacing in the "as found" condition. It is advisable to clean the pendulum slider before testing again.

Proposed technique (dry contamination)

Because the contaminant tends to be swept away on the initial swings, a modified procedure was adopted in research on this condition (Lemon *et al*, 2001).

1   Three conditioning swings are made on a pristine, uncontaminated floor surface.

2   Dry contaminant is applied.

3   Five test swings are carried out.

Where there is little variation in the test swing results then the representative slip resistance value (SRV) is taken to be their arithmetic mean. Should there be significant variation, the first reading alone is likely to be the most appropriate. Research has shown that in situations where the level of dry contamination is high, this may affect the readings as a result of drag on the pendulum head caused by contact with compressed contaminant during the initial swings. In such cases the test surface should be wiped between readings, but in other cases this should not be done.

3.3   SURFACE ROUGHNESS MEASUREMENT

Commercially available micro-roughness meters can be used to measure the roughness parameter Rz (Figure 3.3). This has also been known in the past as RzDIN and R_{tm}. Rz is a measure of the total surface roughness, calculated as the mean of several peak-to-valley measurements. Although other industries use the various measurements that can be made, eg average peak height (R_{pm}), HSE reports the use of Rz as being quick, simple and a good guide to floor surface slip resistance (HSE, 2004a).

The shape and spacing of the roughness peaks is also known to influence slip resistance. HSE/HSL intend to further their research in this area in order to understand more fully the inter-relationship of other related surface parameters.

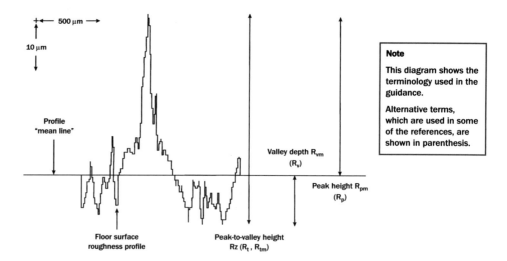

Figure 3.3   *Idealised surface micro-profile: roughness terminology*

Surface roughness is useful to monitor change, and to obtain data in locations and on surfaces and profiles where it can be difficult to utilise the pendulum test effectively.

It is important that the correct surface roughness parameter is used in comparing different floorings, and in utilising the SAT and Slip Potential Model.

When using the meter, small vertical movements of the stylus are detected as it passes over peaks and troughs on the material surface, which are then electronically converted into a measure of surface texture.

Although the use of portable, commercially available roughness meters for the assessment of floor surface slipperiness is increasing, they are unsuited to some common floor types, including carpet and unusually rough or undulating floors (see Section 3.5). As such, roughness measurements should only be used as a guide and should not be used as a sole indicator of the slip potential of flooring materials. Roughness measurements may be used to monitor changes over time of surface characteristics, however.

HSE does not endorse specific meters but its website, <www.hsesat.info>, lists a number of suitable options.

Figure 3.4   *Typical roughness meters*

To achieve a representative figure for surface roughness, it is recommended that 10 readings be taken, in a variety of orientations. These are then averaged to obtain the representative Rz value for the surface.

It is important to ensure that proper contact is made between floor and instrument over the test length and to have regard to uneven or rough floor surfaces that might generate unrealistic Rz values.

Floor roughness values for water-wet* conditions have been categorised as shown below in Table 3.3.

Table 3.3 *Surface roughness related to slip potential (Table 2 in HSE, 2004a)*

Rz surface roughness (microns)	Potential for slip†
Below 10	High
10–20	Moderate
20 or above	Low

†   The risk categories are quantified in Section 4.2.1

In the presence of other contaminants differing levels of surface roughness will be required to lower the slip potential. As a general rule, the level of surface roughness required is related to the viscosity (or thickness) of the contaminant, as discussed in Chapter 5.

The results are also related to "normal" use by pedestrians, meaning that they take place on horizontal surfaces with no personal encumbrances such as loads to carry or items to push or pull. In the latter circumstances, higher requirements are demanded.

It has been shown that necessary roughness levels for soft floor surfaces (vinyl, linoleum) are slightly less than those for harder floor surfaces (tile, ceramics) for the same degree of risk (Hughes and James, 1994).

3.4   HSL RAMP TEST

This test, which HSE has developed more recently from work by Rapra Technology Ltd, is derived from tests described in two German standards DIN 51097:1992 and DIN 51130:2004. The ramp test allows a laboratory assessment of slipperiness through the use of a test subject who walks backwards and forwards over contaminated flooring at ever increasing angles of inclination.

These DIN standards respectively utilise barefoot walkers with soap solution as a contaminant, and walkers wearing safety boots with oil as a contaminant. HSE has reservations about both tests, as neither uses contaminants commonly found in workplaces, nor do they allow for a range of footwear. In addition the classification scheme used in the DIN standards has caused some confusion and misapplication of floor surfaces around the UK (Bailey, 2005) – see Appendix 5.

---

*   This is described in Chapter 5; it is considered to be a realistic base measurement.

Accordingly, HSL has developed a modified version of the DIN ramp test that addresses these issues. It is known as the HSL ramp test and the contamination is provided by clean water sprayed over the floor surface at a rate of 6 litres/minute. It is the use of water that is the prime differential from the DIN test itself. HSE has adopted this methodology because it allows comparison with site flooring in typical water-wet conditions and with data generated by the standard pendulum test. By using standard footwear (Four-S) the test may be used to assess the slipperiness of flooring under known combinations of flooring and contamination. Alternatively, it may be used to assess footwear materials by using standard flooring surfaces and controlled contamination. Hence the advantages of the ramp test are that it allows:

- bespoke combinations to be assessed
- barefoot testing to be undertaken
- useful information in respect of profiled and open-grid surfaces.

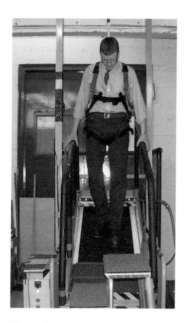

Figure 3.5   *HSL ramp test*

3.5   APPLICATION OF PREFERRED TESTS

Generally it will be possible to obtain parameters from both the pendulum test and the surface roughness meter. However there are circumstances where the nature of the surface or the *in-situ* location prevents their use. Typical examples of these exceptions are listed below.

Profiled surfaces (Section 10.4)

The ability to utilise either test is dependent upon the nature of the profile and the skill of the operator.

Flexible and rigid surfaces (Sections 10.6 and 10.7)

It is not possible to use the roughness meter on carpets or surfaces with large undulations.

Accessories (Section 10.8), stairs and nosings

The ability to utilise either test *in situ* is dependent upon the space available. It is normally not possible to use the pendulum easily on stairs and nosings.

3.6 OTHER TECHNIQUES

Other techniques, to which reference is made in some of the literature, are described in Appendix 5 and include:

- sled tests
- German DIN 51130 ramp test, mentioned in Section 3.4 above
- SATRA friction tester.

Although some promising developments have been made, HSE has questioned the validity of some of these methods, and the sled tests in particular, believing them to produce potentially misleading data (see HSE solicitor's correspondence (Rowland, 1995) and Section 5.1 of *Slip resistance of hard flooring* (TTA, 2002), which acknowledges these shortcomings).

The reasons behind HSE/HSL's concerns are:

- existing sled tests do not replicate the hydrodynamic squeeze film developed in real slip accidents
- results obtained from some sled tests are contrary to experience, ie that wet surfaces should be more hazardous than dry ones
- they do not correlate with the pendulum test, which HSE and HSL believe does generate valid data.

4 Selection and management of walking surfaces

> ### KEY POINTS
>
> ◆ *The standards of walking surfaces are governed by statutory requirements, supplemented by good industry practice.*
>
> ◆ *The selection process involves a holistic approach taking account of a number of established influences.*
>
> ◆ *The client and other key parties need to be involved at an early stage.*
>
> ◆ *The slip potential model is the recommended means of selecting an appropriate surface.*
>
> ◆ *The Slips Assessment Tool is a useful supplementary tool for initial design and comparative assessments.*
>
> ◆ *Once a selection has been made it is essential to ensure it is correctly specified and the required maintenance regime is recorded.*
>
>
>
> Courtesy S G Spark

4.1 INTRODUCTION

This chapter describes the process of selection and management of walking surfaces such that the surface is suitable for the purpose for which it is used.*

The Workplace (Health, Safety and Welfare) Regulations 1992 (as noted in Appendix 2) require that "...the floor, or surface of the traffic route, shall have no hole or slope, or be uneven or slippery, so as, in each case, to expose any person to a risk to his health or safety" (Regulation 12 (2)(a)). This demands a holistic approach, utilising the slip potential model, which takes account of all the matters outlined in Section 2.2 and the data provided in other sections of this guide. This is illustrated on the flow charts, Figures 4.1 and 4.2, supplemented by the tables and text within this chapter.

---

* Regulation 12 of the Workplace (Health, Safety and Welfare) Regulations 1992.

In summary, the law requires that:

- walking surfaces be safe and with an effective means of drainage where necessary
- contamination be avoided so far as reasonably practical.

The first bullet point is an absolute requirement – the test of reasonable practicability does not apply, although the courts have allowed some limited leeway in specific circumstances (see Section 9.9.1).

These requirements apply to a "workplace", a definition that extends to all areas that relate to an employer's undertaking, typically as shown in Figure 1.2.

There are two situations to consider:

- new walking surfaces where the initial choice and design of surface should be made with prevention of slipping in mind
- existing surfaces where, if the surface is deficient ie it has or can lead to a low level of resistance to slipping, some form of remedial or improvement treatment may be required to the surface itself.

In both these cases however the management of the surface (eg how it is maintained or cleaned) will play an important co-existent role in achieving an acceptable solution. The Slip Potential Model, considered below, also takes an inclusive approach.

4.2 THE SELECTION PROCESS

4.2.1 Slip Potential Model: general issues

The Slip Potential Model (Section 2.2) provides the framework for the specifier to give due consideration to all the common factors that may influence the selection, refurbishment or maintenance of a safe walking surface.

The successful use of the model depends upon four elements:

- competent analysis of the component factors
- application of the risk hierarchy demanded by the Management of Health and Safety at Work Regulations 1999 (see Appendix 2), where permitted
- a professional judgement
- ensuring that actions are taken at the appropriate time in any new or remedial scheme.

The process is shown on the following figures: these should be read in conjunction with the accompanying notes and the referenced tables. Figure 4.1 relates to new walking surfaces, Figure 4.2 to existing walking surfaces.

Notwithstanding the process outlined in Section 4.2.3, the following minimum standards are likely to be applicable to most walking surface surfaces, consistent with a low risk of slipping. The SRV is the key parameter.

Table 4.1   *Minimum values of slip resistance parameters*

Parameter	Value	Comment
Slip resistance value (SRV) (see Section 3.2)	36	Under worst expected contamination conditions (but see Table 4.2 and Section 4.2.2 below)
Roughness (Rz) (see Section 3.3)	20 μm*	(From Tables 3.3 and 5.2.) Under contamination; note however that the required minimum will rise with the viscosity of the contaminant as indicated in Chapter 5.

*   Note that in the food processing industry a minimum of 30 μm is recommended (HSE, 1996b).

Some clients and others may wish to identify the quantified risk related to the qualitative descriptions used in this guide. Work by BRE (Pye and Harrison, 2003) provides the data set out in Table 4.2.

Table 4.2   *SRV required for various risks of slipping (from Pye and Harrison, 2003)*

Risk 1 in:	Minimum SRV (see also Section 4.2.2)	Qualitative categories (from Table 3.1)
1 000 000	36	Low
100 000	34	Medium
10 000	29	Medium
200	27	Medium
20, ie 5 per cent	24	High

BRE qualifies this data by indicating that it was obtained from a small sample size, using fit persons at a moderate walking pace. The figures relate to a normal walking situation with no pushing or pulling or turning tight corners, and utilising a horizontal surface. If these conditions are not met, the required SRV for the same risk of slipping will be higher. Note that the SRV values given in Table 4.2 are "installed" figures. To compensate for degradation of SRV during installation and a high density of wear, some specifiers stipulate an increased figure.[†]

4.2.2   Sloping surfaces

Sloping surfaces may result from:

- a need to create a drainage path (see Section 5.10)
- to connect areas at different levels.

Slopes between 1 in 100 and 1 in 40 are generally recommended. A slope greater than 1 in 20 is regarded as a ramp.[‡] Where possible, slopes between 1 in 20 and 1 in 40 are to be avoided, as they are potentially more dangerous than lesser inclinations but often lack the additional safety features associated with ramps (see also Section 11.8).

The slope of the walking surface needs to be taken into account when considering SRV values. If the risk against slipping is to be maintained at the same value, the recommended value from Table 4.2 will need to be increased to compensate for the slope. The full sequence of such an adjustment is illustrated below.

[†]   London Underground requires an additional 10 per cent buffer and specifies an SRV of 40 (LUL, 2005a).

[‡]   See in particular the Building Regulations Part M.

SRV result from Table 4.2 (say)   36

Thus,

Coefficient of friction (CoF)	0.36
Slope	10 per cent (say)
Tangent of angle	0.10
Adjusted required CoF	0.46
Adjusted required SRV	**46**

It is important to ensure that those involved realise that this figure (which would typically be inserted into a specification or similar document) allows for the slope.

Similarly, the measured on-site or manufacturer's SRV will need to be reduced, in a similar manner, in order to allow for the adverse affect of the slope, and the adjusted value compared to figures in Table 4.2 to establish the risk against slipping.

SRV result from manufacturer
or on-site test (say)   39

Thus,

Coefficient of friction (CoF)	0.39
Slope	10 per cent (say)
Tangent of angle	0.10
Adjusted CoF	0.29
Adjusted effective SRV	**29**

This effective SRV value matches that required for a statistical 1:10 000 chance of slipping, ie a "medium" risk of slipping (Table 4.2). The specifier would need to consider whether this was sufficient, or if it should be increased to provide a "low" risk solution.

4.2.3   Slip Potential Model: selection tables

The figures on the following pages illustrate the decision process involved in specifying walking surfaces.

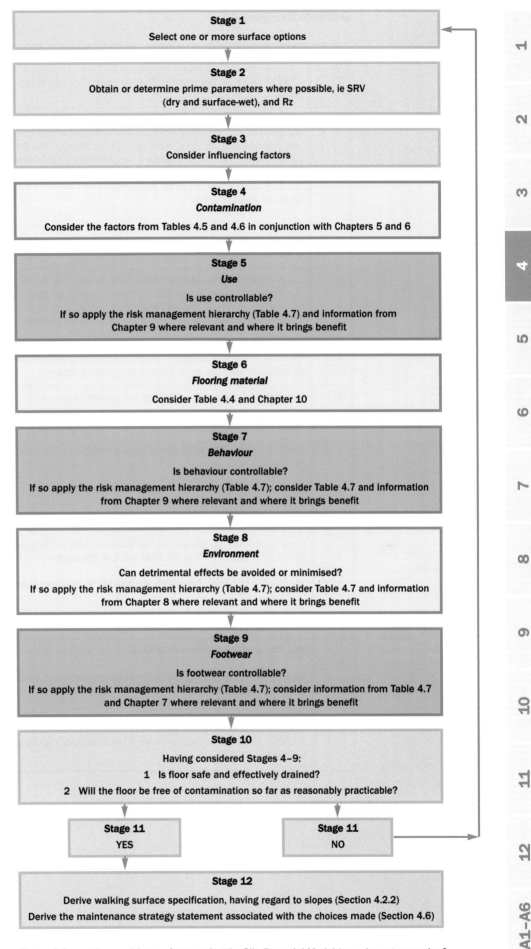

Stage 1

Select one or more surface options

Stage 2

Obtain or determine prime parameters where possible, ie SRV
(dry and surface-wet), and Rz

Stage 3

Consider influencing factors

Stage 4

Contamination

Consider the factors from Tables 4.5 and 4.6 in conjunction with Chapters 5 and 6

Stage 5

Use

Is use controllable?

If so apply the risk management hierarchy (Table 4.7) and information from
Chapter 9 where relevant and where it brings benefit

Stage 6

Flooring material

Consider Table 4.4 and Chapter 10

Stage 7

Behaviour

Is behaviour controllable?

If so apply the risk management hierarchy (Table 4.7); consider Table 4.7 and information
from Chapter 9 where relevant and where it brings benefit

Stage 8

Environment

Can detrimental effects be avoided or minimised?

If so apply the risk management hierarchy (Table 4.7); consider Table 4.7 and information
from Chapter 8 where relevant and where it brings benefit

Stage 9

Footwear

Is footwear controllable?

If so apply the risk management hierarchy (Table 4.7); consider information from Table 4.7
and Chapter 7 where relevant and where it brings benefit

Stage 10

Having considered Stages 4–9:

1 Is floor safe and effectively drained?

2 Will the floor be free of contamination so far as reasonably practicable?

Stage 11

YES

Stage 11

NO

Stage 12

Derive walking surface specification, having regard to slopes (Section 4.2.2)

Derive the maintenance strategy statement associated with the choices made (Section 4.6)

Key

☐ Controllable

▧ Predictable

Figure 4.1   *New walking surfaces: using the Slip Potential Model (see also notes overleaf)*

Guidance for Figure 4.1

Stage	Note
1	It will be usual for the specifier to have more than one option in mind or to hand at this stage.
2	It is essential that the SRV (dry and surface-wet) and Rz are available so that appropriate design decisions may be made based on established parameters and the information provided in this guide be used to best effect. Although the SRV is usually the prime parameter, Rz acts as an important supplementary parameter (see Section 3.1). The SRV value under the worst expected contamination, if not water, should also be obtained when possible. In some cases it may not be possible to obtain both parameters (Section 3.5). Is the surface on a slope? If it is, consider the adjustments indicated in Section 4.2.2.
3	It is necessary to consider the influencing factors, one by one, that feature within the Slip Potential Model (Figure 2.1). Normally they will all apply, but their independent influences will vary. It may be necessary to revisit a decision on one factor having considered the issues with another in order that an appropriate balance may be achieved.
4–9	Consider the issues, using the tables as indicated, the risk management hierarchy, and information from other parts of this guide if required.
10	The specifier now needs to ask the two questions posed. **Is the floor safe and effectively drained** (where required)? There is an absolute obligation to ensure that this is the case, as indicated in Section 4.1. The judgement can be made against the values of SRV and Rz given in the text for various conditions. Remember that a sloping surface will require an enhanced SRV (Section 4.2.2) and that a film of liquid may remain despite the slope. If the answer is "No" (Stage 11), the exercise must be repeated, altering one or more of the parameters, so that the final result is acceptable. **Will the floor be free of contamination so far as reasonably practicable?** If the hierarchy of Table 4.7 has been properly considered, the answer ought to be "Yes" (Stage 11). If not, the exercise must be repeated.
11, 13	Yes. The details can now be finalised. This will involve producing the contract specification and maintenance strategy statement (Section 4.6.2). It is essential to maintain clarity in respect of allowances for slopes (Section 4.2.2).
11, 1	No. All or some of the cycle will have to be repeated until the requirements are satisfied.

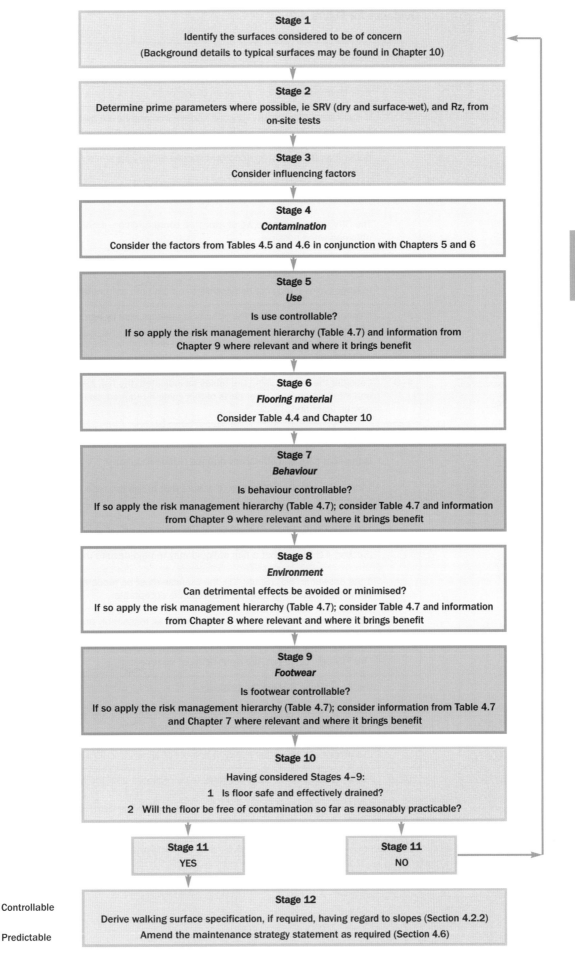

Figure 4.2 *Existing walking surfaces: using the Slip Potential Model (see also notes overleaf)*

Guidance for Figure 4.2

Stage	Note
1	The area of study and/or concern may have arisen from a previous accident, from a maintenance inspection, a change of use or as part of a general upgrade.
2	It is essential that the SRV (dry and surface-wet) and Rz are available so that appropriate design decisions may be made based on established parameters and the information provided in this guide be used to best effect. Although the SRV is usually the prime parameter, Rz acts as an important supplementary parameter (see Section 3.1). Although the SRV is usually the prime parameter, Rz acts as an important supplementary parameter (see Section 3.1). The SRV value under the worst expected contamination, if not water, should also be obtained when possible. Is surface on a slope? If so consider the adjustments indicated in Section 4.2.2. In some cases it may not be possible to obtain both parameters (Section 3.5).
3	It is necessary to consider the influencing factors, one by one, that feature within the Slip Potential Model (Figure 2.1). Normally they will all apply. This will allow an assessment of where the problems may lie and whether the walking surface complies with the law.
4–9	Consider the issues, using the tables as indicated, the risk management hierarchy, and information from other parts of this guide if required, so that improvement actions may be identified..
10	The specifier/manager now needs to ask the two questions posed. **Is the floor safe and effectively drained** (where required)? There is an absolute obligation to ensure that this is the case, as indicated in Section 4.1. The judgement can be made against the values of SRV and Rz given in the text for various conditions. Remember that a sloping surface will require an enhanced SRV (Section 4.2.2) and that a film of liquid may remain despite the slope. If the answer is "No" (Stage 11), the exercise must be repeated, altering one or more of the parameters, so that the final result is acceptable. **Will the floor be free of contamination so far as reasonably practicable?** If the hierarchy of Table 4.7 has been properly considered, the answer ought to be "Yes" (Stage 11). If not, the exercise must be repeated.
11, 13	Yes. The details can now be finalised. This may involve producing the contract specification for any surface improvement, or a new maintenance strategy statement (Section 4.6.2). It is essential to maintain clarity in respect of allowances for slopes (Section 4.2.2).
11, 1	No. All or some of the cycle will have to be repeated until the requirements are satisfied.

The sequence outlined in Figures 4.1 and 4.2 is supported by the data set out in Table 4.3.

Table 4.3   *Examples of Slip Potential Model component issues*

	Factor	Typical variables associated with each factor
1	Contamination (Chapter 5)	Likelihood, nature, visibility, avoidance by user, isolation by management, drainage of fluid contaminants* (Table 4.5)
2	Cleaning (Chapter 6)	Frequency, thoroughness, effectiveness
3	Floor surface (Chapter 10)	SRV and Rz values, profile, wear, maintenance, quality of installation (Table 4.4), inclination of surface (assists drainage, but can increase slip potential)
4	Environment (Chapter 8)	Absolute condition, variability, predictability, lighting , noise, visual distraction
5	Human (pedestrian) factors (Chapter 9)	Public/private use, controlled/uncontrolled use, characteristics of use (slow or fast gait, turning movements), gender, age, disability, training, encumbrances such as luggage
6	Footwear (Chapter 7)	Controlled/uncontrolled, specification, maintenance, age, contamination

In addition to the above issues, specifiers of surface materials will need to also consider the points in Table 4.4.

Table 4.4   *Walking surface attributes*

Attribute	Comment
Cost	A floor surface with a higher SRV may be more expensive than an alternative with a lower SRV, but, with appropriate caution, it might allow a less strictly controlled cleaning regime. This emphasises the whole-life costing approach that should be adopted where possible.
Appearance	A smooth, shiny surface might look impressive, but its low micro-roughness may demand a more frequent cleaning and maintenance regime or one that does not leave any residual liquid on the surface. There may be a significant risk of disruption (eg employee absence or civil action) as a consequence of slips, arising from a failure to ensure appropriate cleaning.
Wear/durability	A harder surface will maintain its surface roughness longer than the softer alternative. Some surfaces improve with wear, and other surfaces degrade with time and usage. Aesthetically pleasing surfaces can be achieved using materials with higher SRV and Rz values. Note: all surfaces will wear but how this affects the slip resistance is unpredictable. The only way to be sure is to measure the SRV over time, using the pendulum test (in conjunction with surface roughness measurements as a useful complementary means of monitoring).
Maintenance	The floor surface should be chosen with the use, any legal requirements for cleanliness and the desired maintenance regime in mind.
Consequences of falls	All floors need to be safe, but where a history of civil action or incidents would seriously damage the occupier's image it may be worthwhile to specify a solution that errs on the cautious side.

* Para 98 of the Workplace (Health, Safety and Welfare) Regulations 1992 requires "effective" drainage where the floor is liable to get wet and to the extent that the wet can be drained off.

Contamination

As noted in Chapter 5, the Workplace (Health, Safety and Welfare) Regulations 1992 place specific emphasis on contamination (which includes water and cleaning residues).

Taking the above into account, the approach in respect of contamination should be as indicated in Table 4.5. For new or refurbished floors it is the specifier who should follow this approach as part of the selection or remedial proposals; for operational situations, the relevant manager will be responsible.

Table 4.5   *Control checklist for contamination-related slip risks (derived from Table 5 of HSE, 1996b)*

1	**Prevent contamination in the first place** *If prevention is not reasonably practicable* ⌐
2	**Prevent contamination becoming deposited on walking surfaces** Preventing contamination from reaching the walking surface has been found to be most effective in practice. *If not reasonably practicable then* ⌐
3	**Limit the effects of contamination** • by immediate treatment of contamination, eg by dilution • by safe cleaning methods minimising wet surfaces • by limiting the area of contamination. The necessary Rz will depend upon the nature of the contaminant (Table 5.2); a minimum SRV of 36 should also be provided under the expected contamination. *If a risk remains* ⌐
4	**Regain the original SRV and surface roughness by cleaning effectively** Follow an effective cleaning regime as indicated by the supplier (this may need verification). Train, supervise, motivate and equip cleaners. Maintain surfaces and drainage to maximise slip resistance. *If this is not enough* ⌐
5	**Increase the specified surface roughness of the proposed or existing surface**

MANAGEMENT OF WALKING SURFACES

General issues

Bearing in mind the absolute requirement for safety and the failings often found in management systems (see Table 4.6), a safer approach would normally be to rely on physical properties, ie a minimum SRV and surface roughness, rather than a cleaning regime to obtain the requisite resistance to slips.

Table 4.6 *Management regimes: potential shortcomings*

Cleaning regimes	Susceptible to lowering of standards through staff turnover, low status of job, complacency, failure to manage.
	Lack of appropriate equipment and training.
Local contamination removal	Susceptible to labour shortage, failure to inform new staff, busyness, complacency and poor safety culture.
Stipulated workplace footwear	Depends upon the nature of the workplace but susceptible to loss of discipline over time (eg use of correct footwear or its maintenance). Usually not possible to rely on the public or visitors utilising the appropriate footwear.
Environment	Susceptible to management not knowing, or forgetting, the relevance of various measures, eg lighting or visual signs, and of the creation over time of distractions. Change of use over time.
General	Failure to ensure warning measures are removed when no longer required (see Figure 4.3). Tendency by some managements to give permanent warnings, eg "danger of slipping" when risk is always low.

Figure 4.3 *Poor management – "Wet Floor" sign chained to a handrail for long periods (courtesy HSE)*

Notwithstanding the shortcomings detailed in Table 4.6, other factors relating to the surface itself, such as cost or appearance, may be of such significance that a solution relying on management regimes may be the preferred solution. In this case it is important that the assumptions made by the specifier are transmitted to, and discussed with, those who will be managing the facility (see Section 4.6.2).

Hierarchy of risk management

The specifier has the obligation to follow the hierarchy of risk management* when considering the influencing factors. This is best illustrated through the use of the acronym ERIC (Table 4.7). The checklist in Table 4.5 is consistent with this approach.

Table 4.7   *Hierarchy of hazard and risk management*

Sequence (ERIC)	Explanation and example action (actions are qualified by "so far as reasonably practicable")
Eliminate	Is it possible to eliminate the hazard? If so, this must be the first choice. Example: clean floors out of hours so as to avoid any wet contamination from this source during the period of use.
Reduce	If the hazard cannot be eliminated, can it be reduced in any way? Example: erect barriers, dilute wet contamination, provide signs at change of floor surface, instruct use of safety footwear.
Inform	Having taken the steps outlined above, is there any residual information that needs to be passed on to the occupier/maintainer (through the health and safety file if there is one, or through operating manuals)? Example: "Any spillage in the food preparation area should be removed and the floor cleaned *and dried* immediately† to maintain its slip resistance".
Control	These are the measures that need to be put in place by those in charge of the premises, so that the assumptions of the specifiers are met. Example: a cleaning protocol; staff training; maintenance of lighting; checking surface roughness; replacement/refurbishment of the surface after a specified time/deterioration.

It will be found from the actions stipulated in Figures 4.1 or 4.2 that in many cases the contribution of some of these hazards to the Slip Potential Model may be eliminated, or significantly reduced, by careful design or operational management. This will reduce the number of variables around which the final decision is made.

4.4   DESIGN AND MANAGEMENT SUMMARY

Having gone through the process illustrated in Figures 4.1 and 4.2, and having included each of the six attributes of the Slip Potential Model where relevant, a professional judgement has to be made on the facts presented. Such a decision is likely also to be a cost-effective solution. There is no "right" answer in weighing up the various options. The final decision relies on the judgement of a competent person.

The timing of this assessment exercise is important. On many projects (new-build or refurbishment) the initial cost plan will have a sum included for walking surfaces, but this may have been based on historical data and ignore the wider holistic approach described above. Once fixed it is usually very difficult to increase the figure at a later date if it is found to be inadequate.

The actual choice and detailing of the floor surface is often left until late in the project when the flooring sub-contractor is appointed or the work package is tendered. At this stage it is usually too late to start discussions on all the factors outlined above. These two constraints prevent full utilisation of the Slip Potential Model, which makes it particularly important to engage with the client and all interested parties at an early stage, and to consider fully the whole-life cost implications.

---

*   Regulation 3 of the Management of Health and Safety at Work Regulations 1999

†   A requirement of para 95 of the guidance to the Workplace (Health, Safety and Welfare) Regulations 1992

The key to a successful and a legally compliant choice of floor surface is therefore to:

- consider the parameters early in the design stage
- discuss flooring surface options with specialist flooring contractors if the specifier does not have sufficient experience
- discuss with the client and, if known, those who will be maintaining the facility or, if not, with those having experience of this work
- ensure that manufacturers' data relate to proven test methodologies
- ensure there is an informed interpretation of manufacturers' data
- derive a solution based on the Slip Potential Model.

4.5 USING THE SLIPS ASSESSMENT TOOL

Although the SAT (described in Section 2.3) should not be used for primary selection purposes (on which final design solutions are based) it may be helpful in providing approximate comparisons between alternative solutions in the preliminary stages. The comparisons may take account of differing cleaning regimes, wear on the material surface over time, and other factors. It could also be used as a monitoring tool on the chosen solution, providing data over time to indicate loss of resistance to slip (ie as a maintenance management tool).

4.6 SPECIFICATION OF REQUIREMENTS

Once the Slip Potential Model analysis is complete, it will be necessary to ensure this is delivered through a combination of material specification and clear maintenance requirements.

4.6.1 Material specification

This will be part of the construction or refurbishment contract.

The requirements may feature as either an end product (prescriptive specification) or a performance specification. In both cases, it is important that the prime surface parameters are in terms of SRV (dry and surface-wet) and Rz values, where these are appropriate to the chosen material. When specifying required SRV values it is essential to have due regard to slopes (Section 4.2.2).

It is also essential that no adverse change in characteristics of the walking surface occurs during the installation process that might reduce the slip resistance, unless this is specifically allowed for in the design and agreed by the facility owner and/or operator, and the recommended maintenance regime is modified to suit.

4.6.2 Maintenance requirements

The condition of a walking surface should be monitored through the life of the asset. Regular asset condition assessments should monitor SRV such that the replacement of the floor (or restoration of SRV) can be programmed into maintenance and refurbishment schedules.

It is recommended that the sequence of discussion and information flow in respect of maintenance requirements follows the philosophy set out in the CIRIA book *Safe access for maintenance and repair* (Iddon and Carpenter, 2004). The principle of early discussion and involvement of all interested parties, set out clearly in the CIRIA guidance, is applicable to the design of walking surfaces. This should result in the specifier producing a *maintenance philosophy statement* that has been agreed with the facility owner and/or operator.

This statement should include the following elements.

Physical characteristics

- The slip resistance parameters (SRV (dry and surface-wet), Rz)
- the anticipated life of the floor product and the limiting characteristics, eg design values of SRV and Rz
- the locations where it is anticipated the surface will require treatment after a period.

It may also be desirable to provide the design SRV under the worst contamination conditions envisaged if these differ from the surface-wet value.

Contamination issues

- The anticipated hazards, eg specific and general contamination
- the assumed cleaning regime and associated maintenance.

Other influences

- The design assumptions regarding other influencing factors, including environment, footwear, use and behaviour (where relevant).

Management

- The anticipated maintenance regime (its type, frequency and methodology)
- day-to-day management issues, eg drying and cleaning of door mats
- re-lamping schedules to ensure the correct luminance.

This maintenance philosophy statement provides important information to those who will have responsibilities for the walking surfaces during their lifespan. The statements made do not prevent alternative approaches being utilised for cleaning and management, so long as they create an equally safe end result.

KEY POINTS

◆   *Contamination may be wet or dry and includes water.*

◆   *Research has established the key role of contamination in slip accidents.*

◆   *Dynamic squeeze film theory is the means by which the action of contamination is explained.*

◆   *Different types of wet and dry contamination result in varying SRV values.*

◆   *The cleaning of surfaces to remove contamination is important.*

◆   *The use of an appropriate maintenance regime to control contamination is essential.*

Courtesy HSE

5.1   INTRODUCTION

Contamination is defined as the presence of any substance on the floor surface, in either a wet or dry state, which has the potential to impair the slip resistance. Hence, in this definition, contamination includes clean water and general dust.

The presence of such substances may be unintentional, such as a spillage, or be intentional, eg a cleaning fluid, or wash water. Similarly, the presence of contamination will range from being avoidable (subject to care and attention), through to wholly unavoidable (for example wetting adjacent to a swimming pool). In between these extremes the ability to avoid contamination will be dependent upon the efficacy of the risk management system implemented by those in charge of the premises.

Contamination may occur in any circumstance, but certain workplaces have particularly poor records, particularly those of the food, catering and hospitality industries and manufacturing; guidance is available directed specifically at these industries (HSE, 2001; HSE, 1996a). It has been confirmed that contamination, ranging from water to heavy soiling, is the main source of slipping accidents on level flooring (Taylor, 1998).

Examples of contamination are given below in Table 5.1.

Table 5.1 *Types of contamination*

Type	Example source
Rain	External areas Transmitted internally from open external doors or from the feet, coats or umbrellas of pedestrians. Leaks through the building envelope.
Water, other fluids	From spillages, leaks, cleaning, drinks machines, production processes.
Cleaning fluids	Arising from the floor management regime and from departures from agreed cleaning regimes.
Process or by-product contaminants	Could be any dry or wet material.
Body fluids and waste	Blood, vomit (human), bird droppings, animal excrement.
Condensation	From variations in temperature or humidity (although the effect is no different than "water", it may be less anticipated by the pedestrian and is virtually invisible). See also Section 8.5.
Dusts	Natural or process-derived, bagged or stored materials.
Debris	Soil, food residues, bags, paper, polythene and other sheet materials, sawdust.

Some of the above examples of contamination may occur on both external and internal walking surfaces (see Figure 1.2) and there is the potential for cross-contamination between areas. Some sources of contamination, eg bird droppings, generally associated with external surfaces, may occur internally, for example on railway station concourses.

The results and data given in this chapter relate to "normal" walking, ie free from carrying, pushing or pulling. In the latter circumstances the slip resistance needed will be greater; the actual requirement will depend upon the specific situation.

5.2 STATUTORY ASPECTS OF CONTAMINATION

The Workplace (Health, Safety and Welfare) Regulations 1992 place particular emphasis upon contamination (see Appendix 2). In essence, the approved code of practice (ACOP) that accompanies the Regulations states that:

- areas likely to get wet or that are subject to spillages should be of a type that does not become unduly slippery. A slip-resistant coating should be applied where necessary. Surfaces near points of danger, eg machinery, "should be slip-resistant" and "kept free from slippery substances or loose materials" (para 93)

- action should be taken to avoid spillage in the first instance (para 94). Where spillage does occur, and where it is likely to be a slipping hazard, immediate steps must be taken to fence it off, mop it up,* or cover it with absorbent granules.

These actions in respect of the contamination are to be achieved so far as reasonably practicable (Regulation 12 (3)), ie the Regulations allow a lesser standard than is required for the floor surface itself. An example of case law is given in Section 6.6.

---

* This is the wording of the ACOP. Research has shown however that even small amounts of residual water on some surfaces can create a slip hazard.

5.3 RESEARCH

A concerted research effort, primarily by HSL, has led to a significant body of knowledge in respect of the effects of contamination on the slip resistance of floor surfaces. This has mostly been directed at wet contamination, as this accounts for the largest proportion of accidents (>90 per cent of accidents in the food and catering sector). Dry contamination, such as flour, talcum powder, sawdust and lint dust, may also create significant slip risks (Lemon et al, 2001).

The pedestrian Slip Potential Model differentiates between wet and dry contaminated floor surfaces. Generally the comments made in Sections 5.4 and 5.5 relate to wet contamination; dry contamination is discussed in Section 5.6.

5.4 HYDRODYNAMIC SQUEEZE FILM THEORY

An understanding of the hydrodynamics of the wet contamination layer is essential to the derivation of safe flooring in respect of pedestrian slips, as it illustrates and explains the fundamental mechanism of a slip on a contaminated surface.

This theory was first researched in 1985 and relates to the fact that pedestrian slipping involves uncontrolled movement of the heel. Hydrodynamic theory was used to study the kinetics observed when a heel comes into contact with a wet contaminated floor surface.

The research found (Lemon and Griffiths, 1997; Lemon, 2003) that the formation of a hydrodynamic squeeze film between shoe and floor could be responsible for a delay in solid-to-solid contact, thereby significantly reducing frictional properties. The squeeze film was able to withstand direct load, but had no shear resistance (in the direction of plate (heel) movement), a feature that could lead to aquaplaning.

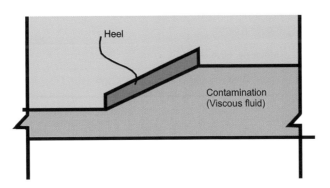

Figure 5.1   *Diagram of a squeeze film*

Subsequent consolidating work (Lemon and Griffiths, 1997) has applied this theory to pedestrian slipping, building on original research by HSL, which had found that the level of floor surface roughness required for solid-to-solid contact increased with the viscosity of the contaminant under study. HSL subsequently confirmed these findings for a greater range of laboratory-controlled contaminants.

The HSL consolidating research findings resulted in two main conclusions.

1 If a fluid of known viscosity is expected to contaminate a floor on which pedestrian traffic is likely, then a surface should be used with a roughness sufficient to break through the squeeze films formed by this contaminant.

2 If it is possible to do so, the viscosity of workplace contaminants should be minimised by substitution or dilution.* This reduces the level of surface roughness required for squeeze film breakthrough in the laboratory and is therefore likely to increase slip resistance in practice.

The roughness levels required to break through hydrodynamic squeeze films formed by contaminants, sufficient to give a pendulum slip resistance (SRV) of 36 (classed as low slip potential – see Table 3.1) are shown in Table 5.2

The minimum roughness figure (Rz) required to allow break-through of squeeze films of clean water has been found to be about 10 microns (Table 4 of Lemon and Griffiths, 1997) but 20 microns is usually taken as a base figure in the presence of workplace contamination.

As a general rule, the level of surface roughness required is related to the viscosity (or thickness) of the contaminant. Although the research (Lemon and Griffiths, 1997) was carried out using research-grade, quality-controlled fluids of known viscosity, the descriptions in Table 5.2 use analogous workplace equivalents.

Table 5.2   *Roughness details (from Table 3, HSE, 2004a; Lemon and Griffiths, 1997)*

Minimum roughness (Rz) (µm)	Contaminant [1]	Viscosity [2] (centipoise (cP))[3]
20	Clean water[4], coffee, soft drinks	1
45	Soap solution, milk	1–5
60	Cooking stock	5–30
70	Motor oil, olive oil	30
Above 70	Gear oil, margarine	30[5]

1   HSE, 2004a

2   Lemon and Griffiths, 1997

3   Where 1 cP= 1 millipascal second (mPa.s)

4   Clean water alone is usually indicated as requiring a roughness value of 10 microns.

5   The research was not able to recommend roughness levels for contaminant viscosities above 30 cP (although some results were obtained), but further observations have allowed some extrapolation of roughness values above 30 cP with confidence.

5.5   REAL WORKPLACE CONTAMINATION

More recent research however (Lemon, 2003) moves away from the technical approach used previously (utilising laboratory-grade materials), so as to make the results more applicable to the real workplace. This specifically analyses a wide range of real workplace contaminants, which are presented as a matrix of floor type against contaminant, and slip potential, in Table 5.3.

---

*   The hierarchy of action as required by the Management of Health and Safety at Work Regulations 1999, and reflected in HSE literature, eg HSG156 (HSE, 1996b). See also Table 4.5.

Table 5.3 *Flooring/contaminant performance matrix (Table 6 from Lemon, 2003)*

☐ = Low potential for slip

☐ = Moderate potential for slip

▨ = High potential for slip

Contaminant	Safety resilient vinyl (Rz 25 μm)	Unfinished quarry tile (Rz 15 μm)	Smooth vitrified tile (Rz 4 μm)
None	Low	Low	Low
Antifreeze/coolant	Moderate	Low	High
Beer	Low	Low	High
Blood	Low	Low	High
Brake fluid	Moderate	Moderate	High
Coffee	Low	Low	High
Cola	Low	Low	High
Cream	Low	High	High
Custard	Low	High	High
Drinking water	Low	Low	Moderate
Furniture polish	High	High	High
Gear oil	High	High	High
Hydraulic fluid	High	High	High
Margarine	Moderate	High	High
Milk	Low	High	High
Motor oil	Low	Moderate	High
Perfume	Moderate	Low	High
Soap	High	Moderate	High
Soap solutions	Low	Low	High
Stock	Low	Low	High
Vodka	Low	Low	High
Yoghurt	Low	Low	High

The original data illustrated in Table 5.3 is qualified by HSE with the following caveats.

1 The contaminants used were of a commercial grade other than where noted. Similar contaminants with non-identical properties may give different results.

2 New flooring samples were used for each test to avoid cross-contamination.

3 Flooring samples were ex-factory and may not replicate aged or as-installed surfaces.

The results allow contaminants to be placed into groups, reflecting decreasing slip resistance, as shown in Table 5.4.

Table 5.4   *Categories of slip resistance*

Group	Typical contaminant	Slipperiness
A	Alcohol-based (vodka, perfume)	Low
B	Water-based (water, coffee, beer, cola)	
C	Moderate viscosity water-based (soap, solution, stock, blood, milk)	
D	Moderate to high viscosity (yoghurt, liquid soap, brake fluid)	
E	Oil/fat-based (motor oil, cream, gear oil, hydraulic fluid)	High

Table 5.4 indicates the relative slip potential of the range of contaminants. The actual slip risk posed by these contaminants in a specific situation will depend on the type of installed walking surface (and the other factors included in the Slip Potential Model).

The researchers recommended further work to build on this knowledge. In particular they noted the novel results, with the potential to disagree with currently accepted theory, whereby low-viscosity alcohol-based contaminants (such as vodka), resulted in lower slipperiness than water-based fluids of higher viscosity.

5.6   DRY CONTAMINANTS

Although dry contamination is less of an issue than wet contamination (in terms of number of accidents caused), it nonetheless demands attention. Dry contamination may take various forms; examples are given in Table 5.5.

Table 5.5   *Examples of dry contaminant*

Natural materials in a manufactured state	From process spillage, eg flour, sugar, salt
Subsidiary materials	Polythene bags, paper Cardboard (perhaps laid over wet contamination to avoid users coming into contact with it, but inadvertently creating another source of contamination)
By-products	Sawdust

It is worth noting that floor mats are a potential cause of pedestrian slips, specifically so when placed on smooth floor surfaces. They need to form part of a managed floor maintenance regime if they are to work as a positive aid against slipping accidents. See Section 11.4 for further discussion on mats.*

---

*   They are also a common cause of trips.

Figure 5.2   *Dry contamination from machine waste on a very smooth power-floated concrete surface*

Two methodologies have been researched for measuring slip, on surfaces subjected to dry contamination (Lemon *et al*, 2001). The methodologies used were the pendulum and the sled (type FSC 2000).

Although HSL has dismissed the sled as unreliable for wet contaminants (as noted in Section 3.2 and Appendix 5), it nevertheless considered the device's applicability for dry contaminants. In the event, the work demonstrated that there was only moderate correlation between the two methods for on-site testing, and poor correlation for the subsequent laboratory testing. The researchers recommended that the pendulum be used alone, for the assessment of the CoF of floor surfaces contaminated with dry contaminants. It was found that it was necessary to modify the pendulum's use from that adopted on wet contaminated surfaces, to reflect the characteristics of dry contamination. The modified technique is also described in Section 3.2.

The results from this research indicate that much more needs to be done before reliable guidelines may be produced. Nevertheless, research into dry contamination concludes that, unlike wet contamination situations, the degree of floor surface roughness (Rz) is not significantly related to slip potential. The researchers comment that it is intuitively logical to assume that the flooring roughness should not affect slipperiness, until the level of surface roughness approaches the dry contaminant grain size.

5.7   CONTAMINATION ON STAIRS

Stairs introduce some issues that are not associated with pedestrian movement on level surfaces. For example, when ascending or descending, the gait of a typical stair-user differs from someone walking on the level.

The effect of contamination on stairs has been researched through several studies (Loo-Morrey, 2003, enclosing BRE, 2001, 2002a, 2002b), which considered in turn, the effect of going size, stair materials and the use of proprietary nosings.

As to be expected, the effect of contamination on stairs is greatest when users are descending and the consequences of falling are so much greater; the effect is more pronounced as the going decreases.

The research also confirmed earlier findings that that the size of going dramatically affects the ease by which users are able to negotiate stairs. Contaminated steps that were found to be hazardous when the going was smaller than 275 mm became usable with larger goings, when test subjects were able to safely negotiate the same contamination on the steps. Stairs are considered further in Section 11.6.

5.8 PROFILED FLOORS

A profiled floor surface is one in which the surface incorporates a deliberate geometric variation, as illustrated in Figure 5.3. Guidance from the HSE (1996b) recommends the use of profiled floors where wet contamination cannot be avoided.

Profiles afford slip resistance where they result in a physical interlock between the profile and the footwear. Where no interlock occurs, so it is possible for the footwear to skate across the profile, then the surface micro-roughness associated with the profile is the controlling feature of the slip resistance in the contaminated condition.

Profiles provide the opportunity for the contamination to collect in discrete areas, while still allowing solid-to-solid contact between shoe and floor, because of the ability of the profiled surface to pool a volume of contaminant before complete submersion.

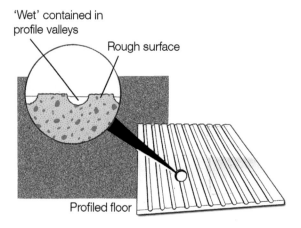

Figure 5.3   *Typical profiled floor surface*

Although this might be expected to improve slip resistance, research (Lemon, Thorpe and Griffiths, 1999a) indicates that significant levels of contamination can remain on the profile plateau, subsequent to drainage. The volume remaining will depend upon the viscosity and tenacity of the contaminant, but the presence of extremely small volumes of contaminant may lead to a sharp decrease in slip resistance. It is in these cases that it may be necessary to rely on the surface roughness characteristics of the profiled tile plateau, rather than other factors such as shape, height, distribution, roughness and waviness, to provide adequate friction in the presence of small amounts of contamination to give satisfactory slip resistance. Profiled surfaces may require specific cleaning methods.

It is essential that existing misconceptions about profiled surfaces – ie they are always safe – are recognised and dispelled.

5.9 MACRO-ROUGH FLOOR SURFACES

Macro-rough floor surfaces are popular in industrial applications where slip problems have either occurred or been identified as possible, because of the inability of the micro-roughness peaks to break through the potentially substantial amounts of contamination.

These macro-rough surfaces are formed by incorporating an aggregate (either quartz, or silicon carbide, carborundum, or occasionally sharp sand) into an adhesive matrix. The size of these aggregates and their projection from the adhesive layer enables sufficient solid-solid contact with footwear for satisfactory slip resistance even where a large pool of contaminant has gathered. This surface characteristic makes technical description difficult because of its irregularity and randomness. To avoid confusing terminology, the term "macro-rough" is used for this type of surface. "Rough" is understood to mean "micro-rough", as found on normal floor surfaces.

5.10 DRAINAGE

It is important to allow for drainage on floors that are likely to be subjected to wet contamination.* Drainage may be achieved by providing an adequate cross-fall to the walking surface. The ability of the contaminant to flow away, and hence reduce the likelihood of slips, will be dependent upon its viscosity and tenacity and the degree of profiling or macro-roughness that is present. The required SRV and Rz will need to account for the cross-fall (Section 4.2.2) and the likely residual contamination.

5.11 CLEANING

The ability to clean floor surfaces is clearly highly relevant to the choice of walking surface. Cleaning of floors leads to improved floor friction once the floor is dry and, as one would expect, lower contaminant levels (Broughton *et al*, 2000). An important piece of research (Holah, 1994), reported on further by HSL (Broughton *et al*, 2000), found that there was no correlation between cleanability and roughness (Rz) on typical floor surfaces used in the food industry. It is linked, however, to the tenacity of the contaminant.

Although cleaning is generally carried out to maintain appearance and functionality in most locations (and for hygiene purposes in specific circumstances), it is not a suitable prime means of controlling slips; the prevention of contamination, so far as reasonably practicable, should be the first aim. However, in the presence of wet contamination the correct choice of surface roughness is more important and effective than relying on a cleaning regime. Cleaning is discussed further in Chapter 6.

5.12 MANAGEMENT OF CONTAMINATION

This chapter demonstrates that management of contamination is key to the maintenance of a walking surface with an adequate resistance to slips. This is reflected in the Workplace (Health, Safety and Welfare) Regulations 1992, which require contamination to be avoided so far as reasonably practicable, and to remove contamination immediately should it occur.

For this to happen, measures are required to control contamination (Table 4.5) and to avoid the potential shortcomings identified in Table 4.6.

---

*   This is required by the ACOP to the Workplace (Health, Safety and Welfare) Regulations 1992, para 98.

5.13 SUMMARY

Coefficient of friction and surface roughness are the key components in reducing the likelihood of slipping in wet contaminated conditions. However, the role of surface roughness is less well understood in the presence of dry contamination.

In general, the magnitude of surface roughness required increases with the viscosity of the contamination.

The use of hydrodynamic squeeze film theory is a valid means of furthering pedestrian slips research and understanding the slip mechanism on wet contaminated surfaces.

Contamination should be avoided wherever possible and spillages promptly removed, and the surface left dry.

Efforts should be made to minimise the viscosity of unavoidable contamination by either substitution or appropriate dilution. Notwithstanding these precautions, regard should be had to the effects of residual contamination, which can be significant on many surfaces.

6 Cleaning

> **KEY POINTS**
>
> ◆ *Poor cleaning regimes can contribute to pedestrian slip accidents.*
>
> ◆ *Effective cleaning should be integral to the workplace management system.*
>
> ◆ *Manufacturers' instructions for cleaning specific flooring products must be followed.*
>
> ◆ *Operatives should be properly trained to use the cleaning equipment supplied.*
>
> ◆ *Surfaces must be thoroughly dried after cleaning.*
>
> ◆ *The use of rough or textured floors is not incompatible with appropriate cleanability.*
>
>
>
> Courtesy HSE

6.1 INTRODUCTION

Cleaning is one of the six factors identified as contributing to pedestrian slip accidents in the Slip Potential Model (see Section 2.2).

The ability to clean floor surfaces is clearly relevant to the choice of surface. Cleaning improves floor friction and, as one would expect, lower contaminant levels (Broughton *et al*, 1997).

Cleaning is carried out partly for hygiene purposes, but the primary driver in many cases is often "premises ambience". It should not be considered the principal means of controlling slips, although, as described below, it can play an important part. Furthermore, cleaning is a significant factor in the occurrence of slips under certain circumstances (Shaw *et al*, 2004).

The cleanability of floor surfaces is important in three areas.

1   The removal of visible surface dirt and contamination allows the floor surface subsequently to be disinfected effectively; this is particularly important in the healthcare sector.

2   The process should allow the original (ie prior to contamination) slip resistance of the surface to be restored

3   The removal of microbial soiling reduces contamination of both products and the environment; this is of particular importance in the food preparation and catering industry (Holah, 1994).

Procedures for cleaning floors include washing, drying, sealing and polishing. The use of lacquers and polishes to achieve particular effects by some cleaning contractors is associated with premises ambience. Where such materials are used they need to be consistent with specified cleaning procedures and other products used, and certainly their use should not compromise slip resistance.

In Chapter 5 it was noted that contamination can impair the slip resistance of a floor. The removal of contamination by cleaning thus becomes vital to the reduction of slip incidents. This applies equally to the routine removal of low levels of contamination, generally dust, derived naturally and to the routine removal of higher levels, which are likely to be use- and/or process-derived, and to the reactive cleaning of spillages.

Some floor areas will, by their use, be subject to heavy and regular contamination. These include commercial kitchens, factory floors, wet areas such as swimming pool surrounds and certain areas within hospitals. The contaminants will vary according to use, although random spillages of other products will always be a risk. While measures to reduce contamination can and should be implemented as part of a risk assessment and mitigation process, there remains a need to deal effectively with the residual substances. External surfaces, particularly around buildings and on station platforms, are prone to contamination with pigeon droppings. Pavements are notably contaminated with discarded chewing gum, a particularly tenacious material (see Section 6.7.2). Both of these contaminants are substances which can cause slips.

Factors associated with slips in relation to cleaning are the method and frequency of cleaning, the type of surface and the cleaning materials, if any, that are used. The rate of reoccurrence of contamination, if definable, is also important; in terms of hierarchy, prevention or reduction of contamination, it may be a more appropriate focus than the cleaning regime itself (see also Table 4.5).

It is important to understand that different procedures and products, while cleaning satisfactorily, may have different impacts on the slip resistance of the floor depending on whether it is used subsequently in wet or dry conditions. Extensive testing of the slip resistance of a variety of surfaces has been carried out under varying conditions of contamination to provide information for designers and building managers (see Table 10.1). This should be incorporated into cleaning schedules to ensure that the materials used are appropriate for the circumstances. Both facilities managers and cleaning operatives need to understand the requirements in the schedules and follow procedures for routine cleaning and isolated spillages alike. Management controls to ensure cleaning contractors are using the appropriate cleaning regimes can be inadequate, particularly where specific cleaning methods are needed for specific areas.

Cleaning is part of the life of a floor and the properties of the floor will change with time, partly as a result of the cleaning process. This emphasises the importance of monitoring the floor's slip resistance during its life to confirm that the appropriate procedures are being implemented. Comparative results may be obtained quickly and simply using a roughness meter to measure Rz. Failure to follow the correct cleaning procedures may not only affect the slip resistance of the surface but can also accelerate its wear and reduce the floor's asset life.

HSE has published an information sheet on floor cleaning, available from its website, which covers this topic in some detail (HSE, 2005a). This further identifies the importance of consultation with the cleaning manager when either new construction or refurbishment of an existing facility is being planned. In the case of the latter, feedback from experience of the original surface will be of value (see also Section 4.4).

Information is also given in *Slips and trips. Guidance for employers on identifying hazards and controlling risks* (HSE, 1996a) and in the Australian *Preventing slips, trips and falls – guidance note* (WorkCover, 1998a). The former contains cleaning advice in Appendix 1; the latter has checklists that cover the following:

- building survey for the prevention of slips and falls
- review of safety management for the prevention of slips and falls (includes section on cleaning).

Both guidance notes contain summary information on relevant legislation (see Appendix 2 to this guide). There are strong similarities between the requirements in the UK and Australia.

Consideration of cleaning methods

- Is the cleaning method appropriate for the floor surface – does it comply with the supplier's recommendations?
- Is there build-up of polish on floors?
- Is there excessive residue of detergent?
- Do employees have to walk on floors wet from washing?
- Are wet floor signs not available or not used correctly?

Source: WorkCover (1998a); Queensland DIR (nd)

The type of surface in relation to absorbency and roughness needs to be considered in determining the most effective cleaning method(s).

Cleaning is identified as a high-priority control measure in the avoidance of slips. A further HSE-funded report notes:

> *Ensure an effective cleaning programme is in place so that spillages/contaminated areas are promptly cleaned (including drying the floor after cleaning and cleaning out of hours whenever possible).*

(Peebles *et al*, 2004)

Slips and trips research has shown that cleaning processes are often poorly thought through; and cleaners are rarely involved in deciding how things are done.

Source: HSE

6.2 CLEANING METHODS

6.2.1 General

The HSL has undertaken a study of different cleaning methods over a wide range of locations to ascertain examples of best practice and identify areas where improvements could be made (Shaw *et al*, 2004). The study concluded that:

- many techniques commonly used are not appropriate for the environment in which they are applied

- the slip resistance of flooring immediately after cleaning could often be improved without requiring a significant outlay of time or money. Information on the various techniques examined is given below

- more research might be carried out into different methods and their efficacy.

Issues relating to the health and safety of those operating the cleaning equipment also need to be considered (WorkCover, 1998b).

6.2.2 Spot cleaning

Using a paper towel or rag to remove small areas of water-based contamination is the cheapest and most effective method of removing water-based spills. It avoids spreading the contamination or increasing the slip risk by mopping a large area. Spot cleaning can be used between scheduled whole-floor cleaning to control contamination. For greasy spills, detergent is required (HSE, 2005a) – see Section 6.2.11, below. Specialist products are available to achieve this kind of cleaning over a wider area in a short time.

6.2.3 Mop

Mopping is generally only effective on smoother floors. Where smooth floors (Rz < 20 microns) are mopped, care must be taken to ensure the floor is left completely dry and that barriers to prevent access to the area are left in place until this has been achieved. Even a well-wrung mop will leave a thin film of water, which is enough to create a slip risk on a smooth floor.

Issues to consider here are how are the mops themselves cleaned and at what frequency. Dirty mops will not be fully effective even where the correct cleaning materials are used. A separate bucket for wringing out dirty water should be used (the two-bucket system). Duplication of function – for example using the same mop to clean and disinfect – may reduce the effectiveness of both processes.

The use of an inappropriate cleaning agent with particular types of mop may itself create a hazard. An example here is the use of oil mops, which if treated with petroleum-base sprays "can turn a safe floor into a hazardous one" (Miller, 1998).

Where the floor is greasy, a simple wash-and-wipe procedure will not fully remove the contaminant and may spread it over a larger area. Contact time with the detergent is needed to loosen the grease. This is known as immersion mopping, in which the detergent is applied in the first stage and mopped up after a soak time in the second stage. In very greasy environments a fat solvent may be more effective than a detergent. The temperature of the solution will also be important to ensure effective dispersion. Rinsing with clean water should be carried out after a detergent has been used, to avoid leaving a detergent residue that may itself cause slips.

Figure 6.1   *Mop and rinse (courtesy HSL)*

Figure 6.2   *Mop and dry (courtesy HSL)*

Machine

Use of a scrubber-drier (S/D) can be an effective way to clean large areas of most kinds of flooring. Certain designs may be more appropriate for particular situations. The machine should be able to operate in the space available and access all areas of the floor effectively. The machine also either needs to be readily transportable to the area to be cleaned or will require secure storage within the vicinity. This has been found to be an issue, for example, in certain underground stations where areas are accessible only via steep or narrow stairs.

The *European Cleaning Journal* (ECJ, 1997) reported on the use of robotic scrubber-driers in the Paris Metro. These were developed by the cleaning contractor under a cleaning contract that had a budget of FFr70 million (approximately €10 million) over 10 years for a programme of research and development of cleaning robots for the Metro. At that stage the technology was not widely applied, but Internet searches suggest that this has changed: several manufacturers offer such products and they are used, for example, at Manchester Airport.

Figure 6.3   *Scrubber-driers (courtesy Tennant Company)*

Rotary mechanical machines fall into three main categories:

* rotary action
* contra-rotating
* cylindrical.

Consideration should be given to the following in selecting or using scrubber-driers:

- the squeegee needs to be wide enough to recover all of the water put down by the scrubber-drier. Single scrubber machines tend to throw water out to one side and may require an asymmetric squeegee to recover this

- the squeegee needs to be well maintained, as any leakage may, for example, leave a smooth floor dangerously wet

- on very rough or profiled surfaces the squeegee may not be malleable enough to allow adequate removal of water from the surface

- very rough or profiled surfaces may also cause premature damage to the machine

- on greasy floors a detergent should be used to remove and hold the oil or grease in the water (HSE, 2005a)

- the operator should be trained in the correct use of the machine, for example using the appropriate level of water for the floor surface, to reduce leaking and water trails (Peebles *et al*, 2004).

Cylindrical scrubbing uses two brushes with contra-rotation. The system is effective on surfaces where:

- there are tiles or recessed grout

- small amounts of debris and high levels of dust are found

- there are irregular features, channels, ridges etc.

Disc scrubbing is effective on very smooth and flat surfaces free from solid debris.

There is considerable variety in both brush material and brush form for cylindrical and disc brushes; selection depends on the intended use in terms of both floor and debris/contamination type. Table 6.1 provides information on the various fill materials available.

Table 6.1 *Fill materials for scrubbing brushes (source Tennant Company)*

Material	Description
Polypropylene	General-purpose scrubbing material for maintaining concrete or coated floors. Can be used indoors or out.
Stiff poly	Similar to poly but has a heavier-gauge bristle that makes it slightly more aggressive.
Non-scuff poly	Lighter-gauge poly for lighter-duty scrubbing conditions and finished floors.
Nylon	Nylon softens when wet, so it is used for gentle scrubbing, mopping or polishing of decorative floors. Will not scratch tile, terrazzo or coated surfaces.
Abrasive bristle	Moisture-resistant nylon filament impregnated with silicon carbide grit. Very effective for grinding away heavy build-ups of soilage and stripping floor finish.
Super abrasive	Similar to standard abrasive, but with larger-diameter filament and heavier grit for additional aggressiveness.
Bassine	Natural fibre, which is a traditional material used for gentle scrubbing of tile, terrazzo or marble.
Union mix	Combination of bassine and another natural fibre called tampico. Most gentle brush fill for light scrubbing or buffing of decorative floors.
Wire	Most aggressive scrubbing brush, used for scraping impacted soilage.

Hose cleaning

High-pressure water cleaning using cold or hot water jets at high velocity can be used to remove dusty or doughy contaminants *provided the floor has adequate surface roughness to cope with the water left behind*. Detergent solutions are often used with this equipment and are needed for greasy contamination.

Mechanical floor scrubber-drier in shopping mall

The problem

A contract floor cleaning company was using rider-operated scrubber-driers to clean a large shopping mall after it had closed for the day. Unfortunately, a pedestrian slipped on an unattended pool of water that had been left by the scrubber-drier. The pedestrian suffered head injuries and spent time recovering in hospital.

On investigation it was found that when the scrubber-drier turned around to begin another run down the mall, the operator raised the blade at the rear of the unit, which deposited water and debris on to the floor.

Rather than cleaning up this water it was left unattended. The floor surface was typical of that found in many shopping centres, being a polished terrazzo tile. When dry, the tile provides adequate slip resistance, but is often very slippery when wet, even with tiny amounts of water.

The solution

New barriers were purchased so that the deposited water was totally enclosed until the floor was dry. The operator had to dismount from the scrubber-drier and set up the barriers, which slightly increased the overall cleaning time. However, there have been no more slipping accidents since this improvement was instituted.

Source: HSE, <www.hse.gov.uk/slips/experience/mall.htm>, accessed 1 Dec 2005

6.2.6   ## *Wet vacuum*

This is effective for cleaning up liquid spills, although moisture or wetness may remain behind afterwards and care is required to ensure that smooth surfaces are left completely dry.

Figure 6.4   *Vacuum (courtesy HSL)*

6.2.7 Dry vacuum

A dry vacuum can be used to remove dusty contaminants, particularly on rougher floors. This is preferable to use of a broom, which is likely to leave a superficial coating of dust on the floor. If the dust creates a health risk a suitable filter must be used.

6.2.8 Sweep

On a smooth floor, sweeping *may* be adequate to remove dry contaminants. There is however a risk that sweeping will raise dust into the air, which then settles back to recontaminate the area. This technique *should not be used* where there are health risks associated with the dust, for example flour or sawdust (Peebles *et al*, 2004).

6.2.9 Scouring pad

Using a scouring pad or brush to agitate the surface, possibly in combination with a detergent, increases the amount of contaminant removed, but leaves detergent *in situ* and could remove carborundum from safety flooring.

6.2.10 Squeegee

The HSL study (Shaw *et al*, 2004) found that use of a squeegee had a negative effect (although in some circumstances it can be useful) in the following situations:

- where the surface roughness is sufficient to allow the floor to be left wet the volume of water is not important and the squeegee is not needed

- where oil or grease is present the squeegee spreads a thin layer of the contaminant over a wider area or forces it into the rough surface. In both cases the outcome is a surface that is harder to clean than it would have been without use of the squeegee.

6.2.11 Detergent

Detergent needs to be used where there is oily or greasy contamination on the floor. Water alone, either warm or cold, will not be effective. The correct concentration of detergent is critical to its effectiveness. The manufacturer's instructions should be followed, as too strong a solution can be as ineffective as one that is too weak: dosing systems can eliminate error. The detergent should be left on the floor for sufficient time to allow effective removal of the grease prior to rinsing – analogous to the soak time required for heavily soiled pots and pans when washing up. Scouring or brushing can increase the effectiveness of detergent. The floor should be rinsed subsequently with clean water to avoid leaving a residue of detergent, which could itself present a risk.

Chemical cleaning may be carried out manually using with a proprietary deep cleaner.

Soap opera

An operatic performance turned into a comedy when a residue of the cleaning agent used was left on the sloping stage. In turn – and in groups when the chorus arrived – the performers were left clinging to a central prop, the only means of avoiding sliding over the edge into the orchestra pit, and to each other. Only the orchestra, sheltered by the stage above, remained unaware of the problem, and its members were clearly puzzled by the mirth of the audience...

Source: BBC Radio 3 broadcast

Table 6.2   *Successful cleaning techniques – easy reference matrix (Shaw et al, 2004)*

Contamination	Micro-roughness Rz (µm)			
	0–20	20–30	30+	Profiled
Water	S/D Detergent Rinse Mop Wet vac Dried	S/D Detergent Rinse Mop Wet vac Dried Hose	S/D Detergent Rinse Dried Hose	S/D Detergent Rinse Dried (dependent on Rz – some profiled floors will not require drying if rinsed with clean water)
Oily/greasy	S/D Detergent Rinse Fat solve Mop Dried	S/D Detergent Rinse Fat solve Mop Scouring pad Hose	S/D Detergent Rinse Fat solve Scouring pad Hose	S/D Detergent Rinse Fat solve Scouring pad Dried – as above
Dry	S/D Mop + dry Vac Sweep	S/D Mop Vac Sweep Hose	S/D Vac Hose	S/D Mop +dry Vac Sweep

6.3   CLEANING DIFFERENT FLOORING MATERIALS

6.3.1   *General*

The specification of the flooring and the particular requirements for its cleaning and maintenance must be made available to those responsible for cleaning post-installation. Relevant information should be included within the operation and maintenance manuals prepared for new construction:

- this should be incorporated within standard cleaning and maintenance procedures
- the procedures should be amended as necessary when the flooring is replaced.

This is reiterated within NHS Estates HTM 61 (NHSE, 1995, currently under revision):

> *...certain cleaning materials can result in shrinkage of pvc flooring resulting in the opening up of joints in tiled floors or the rupture of the welding of sheet pvc floors.*

> *...highly alkaline cleaning agents should not be allowed to build up on ceramic tiled floors as this can cause severe slipping problems and be very difficult to remove. Ceramic tiled floors in particular should always be rinsed with clean water after the cleaning procedure is completed, and allowed to dry.*

The manual notes the importance of obtaining maintenance procedures from the product manufacturer and ensuring these are passed to and understood by the appropriate staff/contractors. (Implementation by the cleaning team is also essential.)

The NHSE manual also contains indicative cleaning regimes (see Table 6.3) for hard finishes, which are defined as:

a homogenous material formed from one or more of the following:

- *thermoplastics*
- *natural stone*
- *clay*
- *sand*
- *cement*
- *thermosetting resins and*
- *soft finishes, defined as "a textile material, woven, stitched or otherwise formed from natural or man-made fibres or both.*

Table 6.3   *Indicative cleaning routines for hard finishes (NHSE, 1995)*

	Daily	Weekly	Periodically
A	Wet scrub		
B	Wet soap with hot water and neutral detergent (disinfectant may be required in some areas)		
C	Damp mop and spot clean	Damp mop with dressing/ water solution and burnish	Machine-strip using hot water and alkaline detergent – rinse to neutral reapply dressing and burnish between coats. In extreme cases at six-weekly intervals, less frequently in others
D	Damp-mop up to three times per day	Damp mop with dressing/ water solution, burnish and spray clean.	As for C in extreme cases at four-weekly intervals, less frequently in others

Table 6.4   *Indicative cleaning routines for soft finishes (NHSE, 1995)*

	Daily	Periodically
E	Suction cleaning and spot cleaning as required	Hot water extraction cleaning

(In Tables 6.3 and 6.4 above the cleaning routines relate to the requirements for different physical and performance characteristics of finishes defined elsewhere in HTM 61, eg Cleaning Routine A is for hard finishes that are jointless, impervious, smooth and anti-static.)

6.3.2   Sheet and tile surfaces

Recommendations for the initial treatment and subsequent maintenance of various forms of flooring in sheet and tile form are provided in BS 6263-2:1991 *Care and maintenance of floor surfaces. Part 2 Code of practice for resilient sheet and tile flooring.* This code of practice looks in general terms also at the care and maintenance of these surfaces in special buildings and particular environments.

Clause 4.1.5 notes that the correct initial treatment of flooring is important and the success of the subsequent maintenance may depend upon its having been carried out correctly. Any seal or other surface treatment provided by the manufacturer as protection will generally be compatible with commercial polishes and *should only be removed when specifically recommended by the manufacturer.*

While the standard contains provides detailed information on the types of flooring, methods of cleaning and cleaning materials, it contains little reference to the slip resistance of the surfaces in various conditions of cleanliness or otherwise. Clause 6.10 covers areas where additional slip resistance is a requirement. This notes that where proprietary types of slip-resistant flooring are installed the manufacturer's instructions should be followed. This agrees with the advice given in general both earlier in this standard and in other publications. Clause 6.10 also points out that high-gloss finishes are often wrongly considered to be hazardous, because to some people they have the appearance of being slippery.

Figure 6.5   *Newly installed terrazzo – highly reflective (courtesy HSE)*

The slip resistance of the flooring is lowered when contaminated by dust, water or materials like oils, greases and similar contaminants.

6.3.3   *Ceramic tiles*

Ceramic floor finishes combine hard-wearing characteristics with the ability to be easily maintained in a sterile condition. This makes them an obvious choice for use on floors (and walls) in food processing and preparation areas.

The following is taken from The Tile Association *Specification for the installation of ceramic tiles in food preparation, treatment and processing areas.*

The Food Safety (General Food Hygiene) Regulations 1995, implementing the requirements of the EC Food Hygiene Directive, require that floors, walls ceilings and surfaces (which come into contact with food) be adequately maintained, easy to clean and, where necessary, easy to disinfect. This requires the use of impervious, non-absorbent, washable and non-toxic materials. Floors should be slip resistant and walls should be smooth up to a height appropriate for the operations.

The Regulations allow the risks to food safety to be taken into account in the application of these requirements. In operations of low risk, materials with some absorption, for example, may be acceptable. This should be confirmed with the food hygiene authority at the design stage. In addition the Materials in Contact with Food Regulations 1987 require that surfaces in contact with food should not transfer any constituent that could either endanger human health or taint the food. Ceramic tiles, installed and maintained in accordance with this specification, provide a durable, attractive, hard-wearing and slip-resistant surface that fulfils all of these requirements.

The TTA technical specification recommends methods for installing ceramic tiles using adhesives that conform to BS EN 12004:2001 and grouts that conform to EN 13888:2002 in order to satisfy the due diligence requirements of food hygiene legislation. This specification may also be appropriate for any type of installation where hygiene is important, such as food storage areas, canteens and restaurants, hospitals and industrial clean areas.

Source: TTA, 2004

The Tile Association has produced guidance *The cleaning of ceramic tiles* (TTA, nd). While this is of use in relation to the methods recommended for removal of different contaminants it has very limited content on the slipperiness of the contaminated or cleaned surface. It is a guide of more general applicability, covering glazed and unglazed ceramic tiles for use on walls and floors. It reiterates the advice given above that manufacturers' guidance should be followed in relation to cleaning products and the compatibility of different materials.

For centuries ceramic tiles have stood the test of time as a durable, attractive and hygienic flooring material. Present-day manufacturing methods are based on precision engineering and today's unglazed ceramic tiles are one of the hardest building materials available.

Independent test laboratories, such as the Campden and Chorleywood Food Research Association, have shown that ceramic tiling conforms to all the relevant legislation, including The Food Safety (General Food Hygiene) Regulations 1995. They say "properly grouted tile surfaces are as cleanable as continuous resin surfaces" – although the issue here is whether in practice they are cleaned as effectively.

Joints grouted in epoxy grout have been shown to have superior bacterial cleanability to vinyl sheet, seamless resin surfaces and even stainless steel. In addition, grouts are now available which contain anti-bacterial additives. Combined with an effective cleaning regime these grouts will not only inhibit bacterial and mould growth in the first place but will kill any small amounts of microbes which may develop.

Source: ECJ, 1997

Section 7 of BS 5385-3:1989 contains recommendations on cleaning and maintaining these elements. Clause 15.5 in Section 3, Design, also provides information on slipperiness. It notes that "...glazed tiles should not be used in areas likely to become wet unless designed to be slip resistant", and also that "Floor surfaces may become slippery in time through the polishing action of traffic".

Figure 6.6 *High-slip-resistant ceramic tiles (55 micron Rz roughness) and rounded nosing (1 micron roughness) (courtesy HSE)*

6.4 EFFECTIVENESS

HSL carried out research on the effects of cleaning/polishing product application on the slip resistance of floor surfaces (HSL, 1999); see Table 6.5.

Table 6.5 *The effects of different polishes and applications on the slip resistance of flooring*

Type of polish	Type of flooring	Discussion
"Standard" polishes applied using manufacturers' recommended methods [not described, wet-mop implied]	Ceramic – vitrified and glazed	The application of standard polishes "does not appear to reproducibly affect slip resistance to any appreciable degree" for either type of ceramic flooring. Polish delaminates rapidly from the smooth, glazed surface, suggesting that the product may have a short active life.
"Standard" polishes applied using manufacturers' recommended methods [not described, wet-mop implied]	Resilient (vinyl) – two types of heavy-duty contract vinyl with different ex-factory textures	The application of standard polishes to vinyl floors resulted in consistent and marked decreases in roughness and CoF. No delamination of the polish layers was observed during testing, suggesting that these products may have significant service lives when applied to vinyl flooring.
"Anti-slip" polishes that claimed to enhance slip resistance in wet conditions	Ceramic – vitrified and glazed	The application of anti-slip polishes "does not appear to reproducibly affect slip resistance to any appreciable degree" for either type of ceramic flooring. Polish delaminates rapidly from the smooth, glazed surface, suggesting that the product may have a short active life.
Standard polish applied by rotary buffer	Ex-factory smooth vinyl	A rotary buffer was used to apply six layers of polish, with testing of the floor CoF and roughness before polishing and after the application of three and six layers. Both wet and dry CoF and Rz roughness decrease as each coat of polish is applied. Use of a rotary buffer results in far thinner polish layers than the use of standard techniques, but repeated application leads to a detrimental build-up of polish layers.

The research concludes that a more detailed study of the effects of the products available, grouped by active ingredient(s), would be of use.

CLEANING AND DRYING

The process of cleaning should not itself introduce an additional slip risk.

The need for drying after cleaning comes from the risk of slip associated with water or other fluids on a surface. "Squeeze film theory [see also Section 5.3] predicts that a film of liquid may be formed between a person's heel and a wet floor surface in certain conditions, which may lead to an uncontrolled slip" (Lemon and Griffiths, 1997). Smooth floors left damp by a mop after cleaning may be very slippery: areas need to be effectively isolated during cleaning, and thoroughly dried as part of the process. The use of physical barriers is recommended rather than relying on warning signs alone, as the latter are frequently ignored (Shaw et al, 2004). This falls within the hazard management process described in Section 4.1. Cleaning should be undertaken in sections where possible to retain a dry path through the area being covered.

Cleaners may themselves be exposed to slip risks during the course of their work. Controls should be implemented to ensure these are minimised, for example by the provision of appropriate slip-resistant footwear.

Figure 6.7 *Warning signs displayed*

There is detailed guidance on cleaning in NHS documentation. In this organisation, cleaning is an important part of infection control and decontamination, as well as a means of preventing slips by a population that may be weak, unsteady, disabled or a combination of these. Cleaning is undertaken against a background of concern over resistant strains of infection that are appearing in hospitals and which can spread rapidly and dangerously. The NHS Purchasing and Supply Agency has developed specifications for "core" cleaning materials for use within the NHS. These contain appropriate levels of cleaning/disinfection ingredients, but also need to be of the correct formulation for the surfaces on which they are to be used.

For any cleaning substance the factors to be considered are:

- the surface(s) to be cleaned
- the particular equipment required
- are special manufacturer's instructions available.

It is important that any such instructions be supplied at the time of installation. The cleaning material may combat the source of contamination, but it is essential that it does not in the process damage the surface and possibly cause a slip (or trip) hazard.

Use of incorrect materials and/or methods can damage a sound floor, reduce its service life and create a greater risk of slips (and trips). While ease of cleaning is an obvious factor in determining the process, the process itself needs to be appropriate for the surface in question.

Where mats are used to reduce contamination of floor surfaces it is essential that both matting and mat wells are themselves cleaned and properly dried regularly and effectively.

Matting still wet when shopping centre opened to the public

The new owners of a shopping centre in the Midlands wanted to cut costs.

They decided that their overnight cleaning operation could be provided more cheaply if carried out early in the morning, before the shops opened.

Following the change several slipping accidents occurred near to the entrance of the shopping centre.

It transpired that the entrance mats, which had been cleaned using a wet cleaning process, were still damp when customers first entered the building. Customers, unaware that the mats were damp, stepped on to the smooth floor with wet shoes.

Research has shown that tiny amounts of liquid on some smooth floors can have a dramatic effect on the floor slipperiness.

The owners quickly reinstated the overnight cleaning regime so that floors and entrance mats were completely dry well before opening. Reported accidents reduced significantly.

Source: HSE, <www.hse.gov.uk/slips/experience/matting.htm>, accessed 1 Dec 2005.
See also <www.sorm.state.tx.us/training2/SlipsTripsFalls/samelevel.htm#wet>

Any spill should be comprehensively cleaned and dried. A superficial clean that only wipes up the contamination and fails to clean or dry the surface subsequently may create more of a risk than the original spill, as the result can be a thin film that is less obvious visually and spread more widely.

Where it is not reasonably practicable to prevent floor contamination, employers need to ensure that the floor/shoe/contaminant combination is suitable to reduce the likelihood of slipping to an acceptably low level. Advice is given in HSG156 (HSE, 1996b) on specifying floors with sufficient surface roughness for the intended use and expected contaminant.

Most smooth floors are slip-resistant when clean and dry. Floors with a SRV below 36 are likely to be slippery. A dry floor may typically have a SRV of about 70, but one which is which is only mop-dry may have a dramatically reduced value, for example 10. This illustrates the negligible improvement obtained by just wiping up spillages but leaving them damp, or by having a dry floor but transferring contamination to it via a wet shoe. Smooth floors are only safe when dry (HSE, 2003a).

Cleaning should be an integral part of the work environment. Employers should manage the risks of slips and trips as they would for any other hazard or aspect of their business, such as product quality (see Chapter 9). As well as being good business practice, it is an obligation under the Workplace (Health, Safety and Welfare) Regulations (see A2).

Deal with any wet or contaminated floors that do occur

- Clean up spillages immediately. This includes spills on any areas on the customer side (if applicable to your business). Don't forget satellite services or self-service areas.

- Ensure there is prompt spotting of contamination.

- Do not leave floors wet after cleaning – clean them to a completely dry finish if possible.

- If "clean-to-dry" is completely impossible then use barriers and "wet floor" warning signs to keep people off the wet area.

- Use cleaning methods that do not spread the problem. Small spillages are often better dealt with using a paper towel instead of a mop that wets the floor.

- Do not use cardboard to soak up spillages, deal with them properly.

- Remove barriers when the floor has dried (and, similarly, only use "wet floor" signs when the floor is wet).

Source: HSE, 2005b

The NHS healthcare cleaning manual (NHSE, 2004) contains detailed method statements for different cleaning tasks on different surfaces.

Several leading legal cases illustrate the liability of occupiers of shops, supermarkets etc and also non-commercial premises. This liability is given substance by the Occupiers' Liability (Scotland) Act 1960 and Occupiers' Liability Acts 1957 and 1984 (though they do not abrogate any common law duty of care). They require that *occupiers take such care as is reasonable in all the circumstances of the case*. Slipping and tripping cases commonly involve accidents caused by foodstuffs or spillages in shops or supermarkets for which the defendant is the responsible occupier. The cases indicate the importance to the occupier of taking "reasonable care in all the circumstances of the case".

Examples of legal cases

Chidgey v Asda Stores Ltd (Unreported, September 30, 2003) (CC)

C slipped on a grape in a supermarket owned by A. C averred that A ought to have had anti-slip mats in place where she fell and that there was no reasonable cleaning and inspection system at the time of the accident. Held, finding A not liable, that A had a reasonable inspection system; it was reasonable to place anti-slip mats in the most vulnerable areas but the area of the accident was not such an area on the ground that it was a "through route".

Laverton v Kiapasha 2002 W.L. 31476475 (CA)

L slipped and sustained injury while walking on the wet tiled floor of K's takeaway premises after drinking with friends. K appealed against a decision awarding damages to L and finding K wholly liable and having breached its duty of care under the Occupiers' Liability Act 1957. Held, appeal allowed; K had taken reasonable care in the circumstances of the case. K had fitted non-slip tiles; it was inevitable that customer would walk in water during wet weather; it was impractical to mop during busy periods and unreasonable to expect that K ensure that the doormat remained in front of the door. Had K been found liable, L would have been 50 per cent contributorily negligent as she had not taken the care reasonably expected from a person when walking on an obviously wet floor.

Source: QBE Insurance

While the above incident refers to an employee within the education system, the majority of people in schools, colleges and universities are not employees but students and pupils. Education sector employers also have responsibilities to protect them from slips and trips. Sites are often busy and crowded. Structured timetables may lead to large numbers of people moving around at the same time, increasing the potential for slip and trip incidents as they focus on reaching the required destination on time at the expense of observing either floor surface contamination or obstacles. Education facilities are also often used both for out-of-hours activities associated with the establishment and by a variety of other groups. Consideration of all these uses is required in determining the appropriate surface and cleaning regime.

Much of the guidance produced relates explicitly or implicitly to internal surfaces. This guide also provides advice in relation to external surfaces within the built environment. These may include car parks, railway station platforms, pavements and estate paths. There are specific issues here associated with the occurrence of ice and snow, which are outside the scope of this book. The requirements for more general cleaning, however, are important. Depending on both the surface and the location there may be a need to remove accumulations of water, mud and spillages of a wide variety of contaminants. Leaves and other natural debris can be major slip hazards, particularly when wet.

A particular issue for some hard surfaces is the growth of algae, which needs to be removed frequently, generally using an appropriate chemical solution. It will recur in shady areas and regular treatment may be needed.

Efflorescence may also be a recurring slip hazard on paved surfaces, emanating from the pavers, the bedding mortar or even the underlying substrate, and should be removed on a regular basis.

6.7 HOW CLEAN IS "CLEAN"

Generally, cleaning relates to the removal of "gross soiling". Where hygiene is an important issue however there are two relevant components to cleaning, namely **gross soiling** and **microbial soiling**.

"Clean" may also be partly a matter of perception. Where ambience is a driving factor, mystery shopper surveys are sometimes employed to test reaction. Capability Scotland, for example, has set up a panel of mystery shoppers to help with its research into how facilities in Scottish towns and cities are meeting legislative requirements on access. The disabled people used as mystery shoppers help test access across a range of services such as town centres, visitor attractions and high street stores.

6.7.1 *Roughness* versus *cleanability*

An important piece of research (Holah, 1994), reported on further by HSL (Broughton *et al*, 1997), found that there was no correlation between cleanability and roughness (Rz) on typical floor surfaces used in the food industry, in either category. It is linked to the tenacity of the contaminant, however.

Given the legal requirement to avoid slippery floors (in the Workplace (Health, Safety and Welfare) Regulations) it is possible that the use of smooth floors in areas that are likely to become wet or contaminated may contravene them. However, in the food industry, where there is a particularly high incidence of slipping, there have been concerns that the use of rough floor surfaces would not comply with food hygiene regulations that require that flooring must be easy to clean.

> *Laboratory measurements of cleanability and floor roughness on a wide variety of food industry flooring samples show that the cleanability of the floor is not linked to surface finish. Cleanability is linked, however, to the tenacity of the contaminant. That which can be removed from a smooth floor can also be easily removed from a rough floor.*
>
> (Holah, 1994)

This would seem to remove any conflict between the requirements for hygiene and the avoidance of slippery surfaces in food production areas.

> *Smooth floors are thought to be easier to clean than rough or textured floors, but it is incorrect to assume that all smooth surfaces are cleanable and all rough surfaces uncleanable.*
>
> (Holah, 1994)

Data from testing allow the selection of flooring that is both microbially cleanable and also of a surface roughness compatible with the required level of slip resistance.

HSL carried out tests on a range of 15 food industry flooring samples. The materials included tiles, resins and concretes.

Roughness measurements were carried out for each sample using a Surtronic 3+ roughness transducer. For each sample, the peak (R_{pm}) and valley (R_{vm}) roughness parameters were measured (see also Section 3.2); the mean of 10 readings was taken as the result for each sample, and the ratio of R_{vm} to R_{tm}, the total roughness parameter for the surface was calculated (Broughton *et al*, 1997) – see also Section 3.3.

While a minimum Rz of 20 microns is recommended (HSE guidance) for flooring in wet conditions to provide adequate slip resistance, this has not always been found to provide an adequate safeguard. It has been shown that the surface structure of the material is important in determining its cleanability. The "asymmetric" floor is defined, where a low R_{pm} with a high R_{vm} indicates a closed structure that affects the cleanability of the surface.

In the same research (Broughton *et al*, 1997), cleanability measurements on the same 15 samples were carried out. There are two stages of cleaning undertaken:

- the removal of visible gross soiling, which is the main source of physical contamination as well as a slipping hazard and a barrier to subsequent disinfection
- the removal of microbial contamination to control product contamination.

The results for a third of the samples, with very different R_{pm} although quite similar total roughness parameters (Rz), were sufficiently similar to suggest that gross cleanability would not vary significantly. This shows that, while rougher floors are required to control slipping hazards under gross contamination conditions, this does not compromise cleanability and there is no link between peak roughness and cleanability of gross contamination.

The testing for microbial soiling showed equally that there is no demonstrated link between the microbial cleaning parameter (assessed as percentage bacteria per cm² remaining on the sample after cleaning under controlled conditions) and R_{pm}, with the **roughest** sample having a cleanability similar to the **smoothest**.

The research led to a number of conclusions in addition to the above, which will be of assistance to those responsible for the selection and specification of flooring:

- microbial cleanability is a function of the contamination rather than the surface finish
- microbial cleanability of a surface cannot be assessed by measurement of its roughness parameters alone
- it cannot be assumed that a floor demonstrating good cleanability characteristics for a particular contaminant would perform equally well for all contaminants: assessment is required for particular products.

It is possible to select a floor that possesses good slip resistance while also having good hygiene characteristics in terms of cleanability.

It should be noted, however, that certain substances have a particular tenacity to macro-rough surfaces and special cleaning procedures may be required for their removal. An obvious, and problematic, example here is chewing gum, which may be removed by the application of WD-40 followed by manual scraping.

Microbiologically clean versus healthcare-acquired infection (HAI)

Concerns have been raised in the past as to whether it is possible to obtain non-slip flooring that also meets hygiene standards. These are of particular relevance in kitchens, hospitals and certain manufacturing facilities, and appear to have been met by materials now available.

Reference is made to the cleaning of non-slip floors in various manuals and cleaning specifications produced within the NHS.

Information has also been obtained on flooring in Australia: flooring materials for kitchens with high slip-resistance are available that meet the recommendations of the Australian Health Hygiene Code.

The *National code for the construction and fitout of food premises* issued by the Australian Institute of Environmental Health is not a state or federal regulation (AIEH, 1993), but many local government organisations require or request compliance with the code for planning approvals.

The code recommends (Part A, clause 2.1) that floors are constructed of materials that are "impervious, non-slip" and (in Clause 2.3) that the floor finish be " ...smooth and even, free from cracks, crevices or surface protrusions that will prevent easy cleaning".

The terms "impervious", "non-slip" and "smooth" are not defined with precision. However, several floor materials with grit-roughened surfaces – typically a PVC flooring containing aluminium oxide grit and a bacteriostat (antibiotics that prevent the growth of bacterial cells) – are marketed with claims that they meet Australian regulations and codes on hygiene and safety. These are now used in many hospital kitchens. A simple sealed epoxy finish might also be considered for such applications.

USE OF THE SLIPS ASSESSMENT TOOL (SAT)

The Slips Assessment Tool, as described in Section 2.3, has been produced following a research project commissioned by the HSE. The SAT is a freely downloadable computer software package that allows an operator to assess the slip potential of pedestrian walkway surfaces. In relation to cleaning, the SAT has the following options to be input to the tool (it also requires information including floor type, surface treatment and potential floor surface contamination sources). For each of these, the appropriate box, and only one within each category shown, is selected to be used in the assessment.

The following is taken from the SAT, showing the information to be input into the assessment in relation to cleaning.

Floor cleaning type

No cleaning attempted		Wet mopped		Mechanical scrubber-drier	
Rotary buffer		Wet mopped and rinsed		Wet-mopped and dried	
Brushed		Water power hose		Wet-mopped, rinsed and dried	
Vacuumed		Squeegee		Other – ineffective	
				Other – effective	

Floor cleaning frequency

Continuously	
At regular intervals	
Once a day	
When it looks dirty	
Seldom	
Never	

Contamination reoccurrence

Soon after cleaning	
Gradually after cleaning	
Very little or no recontamination	

KEY POINTS

◆ *Not all safety footwear is slip-resistant.*

◆ *The properties of the shoe sole are highly relevant to pedestrian slipping.*

◆ *The surface roughness and material hardness of the sole have a significant influence on its frictional characteristics and, therefore, its slip resistance.*

◆ *The wear rate and, to a degree, cleanability of the sole influence the surface roughness levels throughout the life of a shoe sole.*

◆ *Wearing flat shoes that maximise the area of contact with the floor, especially at the heel, can reduce the number of slip injuries considerably.*

◆ *To improve the slip resistance in contaminated conditions, the shoe sole should generally have deep cleating and a well-defined tread pattern.*

◆ *Footwear should fit correctly: slipping is more likely if the wearer's foot moves within the shoe.*

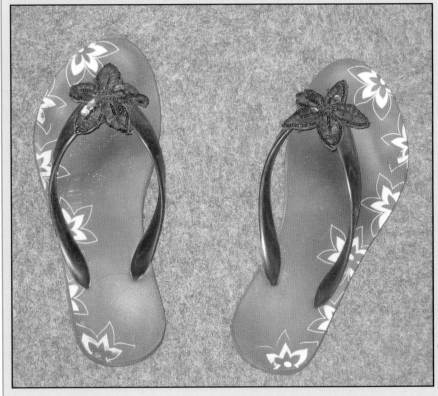

Courtesy Deborah Lazarus

7.1 GENERAL

The properties of a shoe, and specifically the sole, are highly relevant to slip resistance, although control of slip resistance cannot be achieved by assessing the properties of footwear alone. It is the combination of shoe, floor and contaminant that must be controlled in reducing the incidence of slipping. By knowing the properties of different footwear, a reasonable deduction regarding its slip resistance on different floor types in wet and dry conditions can be made.

Although several types of slip-resistant footwear have come on to the market in recent years, it is important to recognise that not all safety footwear is slip-resistant. Users need to be careful in matching the properties of footwear to the likely environments they will encounter. Users should seek up-to-date advice from HSL (<www.hsl.gov.uk/contact-us.htm>). Trialling footwear is the ultimate test of its performance in any environment and will also inform on issues such as durability, comfort and fit.

Footwear will need to be properly cleaned and maintained if it is to remain effective.

Other properties of safety footwear may also be required, for example protection against piercing injuries, and it may be necessary to compromise to achieve more than one safety characteristic.

It is important duty-holders do not dismiss footwear as a control in situations where the public have access. It is often their staff who have a significant proportion of slip accidents and it is perfectly possible to control their footwear.

7.2 SLIPS ASSESSMENT TOOL (SAT)

In assessing the slip resistance of a floor in use, one of the parameters that need to be considered is the type of footwear worn. The HSE Slips Assessment Tool (SAT), which has been developed to assess pedestrian slipping risks on floors, considers six options in the "Footwear" section:

- no control over footwear
- non-safety footwear/other
- standard safety footwear
- specialist anti-slip footwear
- barefoot
- other – cannot identify footwear.

7.3 GUIDANCE FOR EMPLOYERS

Health and safety legislation places duties on employers, owners and landlords to prevent or control slip risks. There are specific legal duties applicable to employers to ensure the health and safety of employees and others who may be affected by their work activities. Guidance is available for employers from the HSE (1996a) to manage slip risks and specific guidance is published for some business sectors, for example food and catering (HSE, 1996b and c), as they report higher than average slip injuries each year.

One of the measures that can be implemented by employers to reduce the slip risk due to employees wearing unsuitable footwear is to ensure that suitable footwear is worn in the workplace. Employers have a duty to provide, free of charge, special protective footwear. The chosen footwear should:

- be appropriate for the task and floor surfaces
- fit properly
- be maintained or renewed as necessary.

Standard safety footwear will be CE-marked: the numbers after the CE marking refer to a notified body that has been involved in the type examination of the product according to the Personal Protective Equipment (PPE) Directive. The requirements and test methods for safety footwear are tested to BS EN 344/BS EN 345/BS EN 346. However, safety shoes are primarily required to protect against sharp or falling objects and the standards do not include test specifications for slip resistance. In fact, there is no agreed European standard test method for the assessment of footwear slip resistance, so any footwear that is marketed as "specialist anti-slip" should be verified independently.

Some sectors, such as the food industry, health services and agriculture, require employees to wear overshoes. These are worn to protect the employees' own shoes from becoming contaminated with liquids and sometimes solids (eg foodstuffs), either of which will also adversely affect their slip resistance.

7.4 CONTACT AREA OF SHOE WITH FLOOR

Slips almost always involve the heel sliding and losing grip on the floor. On stepping forward, the heel makes contact with the floor surface first and normally, as weight is transferred forward, the pressure exerted on the floor increases. Any surface contamination tends to be squeezed out between the sole of the shoe and the floor. Squeeze film theory is described in Section 5.4.

Establishing a "sensible shoe" policy and avoiding fashion footwear in the workplace can reduce the number of slip injuries considerably. Sensible shoes should be flat, to maximise the area of contact with the floor, especially at the heel. By contrast, women's shoes often have small heel areas, especially fashion footwear (eg stilettos), and the type of sole on fashion footwear is often smooth and hard, giving this type of footwear a higher risk of slip. Wearing high heels results in an unnatural body position and can compromise a person's ability to recover after a slip.

When walking on stairs, each step initiates on the toes and ball of the foot rather than the heel for walking on the flat. The likelihood of a slip occurring on stairs is affected by the proportion of the foot that can be placed on the stair tread and the slip resistance properties of the shoe and the stair tread. In ascent, the foot tends to slip forwards first on contact. These slips tend to be very minor in nature and easy for the user to control, without loss of balance. In descent, slips occur when the foot oversteps on to the nosing so that contact is made with only a very small area of the shoe sole. Different types of footwear may also be influential in the likelihood of a slip occurring as it is reported that young females wearing high or semi-high heels experienced relatively high fall rates (Nagata, 1991). It was suggested that the effects of wearing various types of footwear, for example, heeled versus flat, could be investigated on a variable-dimension stair rig.

A BRE client report (BRE, 2002a) recorded observations of two subjects, wearing safety footwear, walking up and down an adjustable stair rig with seven stair tread materials under dry, water-wet and glycerol-contaminated conditions. Under dry conditions, the two users did not slip on any of the seven materials tested, over the going sizes used.

Slips occurred in descent for User 1 when the going was around 250 mm and for User 2, around 325 mm to 300 mm or less, under contaminated conditions (where going sizes varied from 150 mm to 425 mm). The difference between users reflects the difference in their shoe size and gait patterns, with User 1 allowing his feet to turn outwards when descending, while User 2 tended to keep his feet straight throughout the trials. Observations were also reported on glazed ceramic tile steps for a second time but using alternative shoes that had a better resistance to slip on the level. In descent, User 1 started to slip when the going was decreased to 175 mm and User 2 started to slip at 225 mm. This was a marked improvement over the standard boots. Slips occurred during descent as the going decreased and the available friction was low. The user oversteps on to the nosing and the contact area between the stair and the shoe sole is dramatically reduced. During ascent, the ball of the foot is usually placed well forward on the tread and the heel may or may not be placed on to the tread. Similar choices of sole material and cleating pattern should be selected to help protect against stair ascent accidents as walking on a level to try minimising slips. A Health and Safety Laboratory (HSL) study (Loo-Morrey, 2003) suggests that an informed choice of footwear may help reduce accidents on stairs.

7.5 OVERSHOES

A study was carried out by the HSL to investigate the effect of overshoes on pedestrian slipping (Lemon *et al*, 2002). Four types of overshoes were tested on seven flooring materials using potable water as a contaminant. The four overshoes tested were:

- "blue disposable" made from chlorinated polyethylene, popular in the food industry

- "yellow disposable" made from laminated polypropylene/polyethylene composite. Fabric designed to protect against industrial and agricultural chemicals in saturation spray environment

- shoe used in home and site visits, manufactured from injected polyurethane. Lightweight, pull-on lug and textured sole for anti-slip

- reusable overshoe with SFC III™ slip-resistant outsole.

The findings showed that the lightweight disposable plastic overshoes had poor resistance on smooth floors and the blue overshoes in particular were flimsy and prone to tearing under testing. The more substantial overshoes provided reasonable protection against slipping, while the reusable overshoe gave the best anti-slip performance of all.

Despite the comparative success of the "reusable overshoe" in the laboratory-based trial, a real workplace trial revealed that employees were reluctant to wear the overshoes as they did not fit well due to the limited sizes available (small, medium and large). The employer concluded that while the overshoes were good for occasional use, they were not suitable as an alternative to slip-resistant personal protective equipment footwear

7.6 SOLE PATTERN

To improve slip resistance in contaminated conditions, it is important that there is some form of drainage either on the floor surface or on the shoe sole. Profiled floors that incorporate channels or raised sections into the surface of the floor can help slip resistance in wet conditions. The incorporation of profiling on shoe soles is more common, however. These profiles can aid the drainage of contaminants from the contact area between the shoe and floor, thereby increasing slip resistance in contaminated conditions. The design of the sole pattern has to consider the cleating (a cleat in this context is a projecting piece of metal or hard rubber attached to the underside of a shoe to provide traction) and tread pattern to maximise the drainage performance of the shoe. Generally, deeper cleating and a well-defined tread pattern improve the slip resistance on wet surfaces.

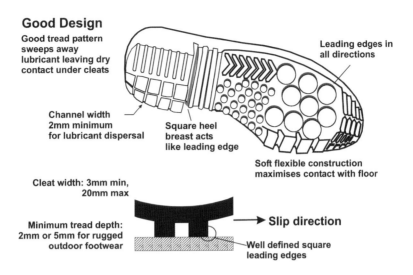

Good Design

Good tread pattern sweeps away lubricant leaving dry contact under cleats

Leading edges in all directions

Channel width 2mm minimum for lubricant dispersal

Square heel breast acts like leading edge

Soft flexible construction maximises contact with floor

Cleat width: 3mm min, 20mm max

Minimum tread depth: 2mm or 5mm for rugged outdoor footwear

Slip direction

Well defined square leading edges

Figure 7.1 *Excerpt from SATRA design guidelines for good slip resistance*

Suggestions for the good design of cleating patterns were developed during a meeting hosted by SATRA on Wednesday 5 March 1997. These ideas appear to have been borne out by footwear with good slip resistance properties that have come to the market in recent years.

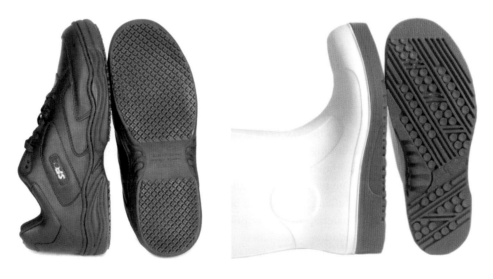

Figure 7.2 *Examples of deep tread profiles (courtesy HSL)*

Different characteristics for the sole pattern are needed to prevent slips in different conditions. On wet surfaces the sole should have a well-defined pattern as more edges will give a firmer grip. The tread will cut through surface liquid and break up the slippery layer under foot. On dry surfaces, it is better to have as much of the sole as possible in contact with the ground, so the pattern on the sole is less important.

7.7 SOLE MATERIAL

Some combinations of shoe sole and flooring materials have been found to be less slippery than others. The relative slip resistance of different shoe sole materials with various floor types (from smooth, through matt to rough surfaces) in water-wet conditions are given in HSG155 (HSE, 1996a). The potential "slipperiness" information is given below.

Table 7.1 *Relative slip resistance of combinations of shoes and floors in water-wet conditions (HSE, 1996a)*

	Floor type	PVC and leather	Urethane and rubbers	Microcellular urethane and rubbers
Smooth	Stainless steel	≈	≈	~
	Polished or glazed ceramic	≈	≈	~
	Finished timber	≈	≈	~
	Smooth resin	≈	≈	~
Matt	Matt ceramic	≈	~	~
	Terrazzo	≈	~	–
	PVC/vinyl	~	~	–
	Concrete	~	–	–
Rough	Paving stones	~	–	–

≈   most slippery
~   less slippery
–   least slippery

When there is no control over footwear, pedestrians will be wearing a variety of sole patterns and sole materials. A percentage of this footwear will be highly unsuitable for use when the floor is contaminated and, as such, the slip risk is high. The risk of slipping in the barefoot state is commonly greater than when the foot is shod. In addition, flooring commonly trafficked in the barefoot state may become more slippery due to the gradual build-up of organic matter. Non-safety footwear comes next in the ranking, as there will be some standardisation in the type of sole pattern and sole materials worn. Because standard safety footwear has a greater degree of standardisation, there is a lower slip risk when compared to the use of non-safety footwear under the same conditions. The lowest slip risk is for specialist anti-slip footwear, as the sole materials will provide the highest coefficient of friction with the various floor surfaces.

The effect of shoe sole material hardness on slip resistance was reported by Hughes and James (1994). They suggested that, for a given level of slip resistance, a higher degree of surface roughness is required for hard sole materials than is required for softer materials. It was found that softer sole materials gave higher friction readings on a dry surface than hard materials.

WEAR AND FIT OF SHOE

Different sole materials and patterns wear at different rates. The wear characteristics of footwear are likely to depend more on the property of a soling material than on the cleating pattern. The rate and manner in which a material wears is important, as the preservation of surface roughness levels throughout the life of a shoe sole is vital in maintaining satisfactory slip resistance in contaminated conditions. Soles made of PVC (a thermoplastic) become smooth and polished with wear and need replacing before they wear smooth. The wear of brand-new leather soles will depend on the nature of the floor surface primarily being walked on. It is possible to get different levels of roughness on a leather sole depending on the initial walking surface (eg they will roughen on street paving). Microcellular polyurethane shoe soles are consistently more slip resistant than their thermoplastic rubber counterparts because the microcellular structure is able to maintain higher surface roughness values than thermoplastic rubber when subjected to wear.

Tread patterns may wear considerably in high-impact areas, such as the trailing edge of the heel, which significantly is the first point of contact between shoe and floor during walking. The occurrence of such wear may reduce slip resistance considerably in contaminated conditions.

New soles may have a skin or film on them from the moulding or forming process. Once this has worn off, the anti-slip performance of the soles will change. Therefore, the footwear will need to be tried over time to assess its slip performance.

It is important that footwear fits correctly. If the wearer's foot moves within the shoe, slipping is more likely because the shoe may not make proper contact with the floor. If a person is comfortable in their footwear, their gait is more likely to be normal rather than placing too much weight or pressure on any particular foot or area of the shoe.

Figure 7.3   *An example of badly worn (and badly fitting) borrowed footwear – the wearer slipped and was badly injured in a slip incident*

The risk of slipping is affected by the length of time that a contaminant remains on the shoe sole. Sometimes the footwear tread pattern can trap a contaminant, thereby reducing its slip resistance. The ease with which a shoe sole can be cleaned also influences its performance. In work sectors wherein liquid and sometimes solid contaminants regularly soil the floor (and hence, the shoe sole) employees are required to wear overshoes so that the covered footwear does not become contaminated. Generally, footwear should be kept free of contamination and the soles should be kept in good repair and renewed when necessary.

SLIP TESTING OF OCCUPATIONAL FOOTWEAR

To reduce the incidence of slipping, a holistic approach needs to be taken so as to control the combination of shoe, floor and contaminant. Each element can be considered in turn, but must be related to the other two to form an overall risk assessment. An informed choice of footwear in the workplace can significantly reduce the level of slip risk. In an HSL report (Loo-Morrey, 2005a) the slip resistance of 27 pairs of occupational footwear was studied. It is intended that the results will form the basis of an information table to aid HSE inspectors in providing advice on appropriate footwear to duty-holders.

Pet food company reduces slip accidents after introducing new footwear

Workers at a pet food factory were experiencing problems from slipping in their work environment.

DIN ramp testing

The work processes meant that it was very difficult to stop contamination – animal produce, fat etc – getting on to the floor. The company had tried many options to reduce slipping risks including anti-slip floors and providing all their workers with footwear as personal protective equipment (PPE).

The company's health and safety managers, in discussion with workers, decided to trial a new style of footwear that had been shown to be highly slip-resistant when tested on the DIN ramp at the Health and Safety Laboratory.

At the end of the seven-month trial, a group of workers using traditional footwear had suffered 15 slip injuries. A similar group using the new footwear had suffered no injuries. The new shoes were more expensive but they lasted longer than the traditional footwear (several pairs lasted three times as long).

Because of the reduction in slip accidents, and the associated savings in lost time etc, the company saved approximately £12 000 in the trial period. The workers also found the new shoes very comfortable

Managers and workers were so impressed with the new footwear that 75 per cent of the 300 workers on site are now wearing them and no-one has had a slip accident to date. The company is also trialling the footwear at some of its other factories around the country and has shared its experience with other companies (small and medium size) and also to a leisure catering and conference provider.

Source: HSE, <www.hse.gov.uk/slips/experience/pet-food.htm>, accessed 1 Dec 2005

8 Environment

KEY POINTS

◆ *The provision of suitable lighting is vital to the maintenance of a safe walking surface.*

◆ *Noise and visual distractions can in certain circumstances increase the risk of slips.*

◆ *Condensation arising from adjacent surfaces can create contamination and hence a risk of slipping.*

Courtesy Deborah Lazarus

8.1 INTRODUCTION

As is recognised in the slip potential model described in Section 2.2, slips may occur not only as a result of the physical state of the floor surface itself, contamination, or footwear, but also as a result of environmental distractions and features, which impact upon the pedestrian. These effects are reviewed below.

8.2 LIGHTING

Lighting is a major contributory factor in pedestrian slips on floor surfaces, especially for the elderly (Cox and O'Sullivan, 1995). The CIBSE *Code for lighting* stresses:

> *The lighting installation must be electrically and mechanically safe and must allow the occupants to use the space safely. These are not only primary objectives but also statutory obligations. It is, therefore, necessary to identify any hazards present…*

(Section 3.1.1, CIBSE, 2002)

Although no specific mention is made of slips, the code gives guidance on good lighting design, which, if followed, is likely to avoid slip problems.

A BRE study summarises the situation succinctly:

> *Visual acuity increases with the level of illuminance. Hazardous situations may thus be created by the inadequate nature of the lighting itself including insufficient light sources, glare, gloom and shadows. For certain tasks (for example, walking down stairs) an adequate level of illuminance is required for users to avoid slips, trips and falls accidents. Low lighting is also likely to increase the likelihood of collision with obstacles or building features.*

(Cox and O'Sullivan, 1995)

Figure 8.1   *Polished sheet vinyl in a supermarket coffee shop area: sunshine through external windows producing glare and reflection on floor surface (courtesy HSE)*

Figure 8.2   *Poor lighting in staircase combined with dirty stairs and additional non-standard nosing*

A change in lighting intensity can have a temporary but negative effect on older people's balance, making them sway more than usual (Easterbrook *et al*, 2001). This also affects the vision of the elderly at night, when falls are commonplace.

The Building Regulations have no requirements for natural or artificial lighting other than emergency lighting for means of escape in the event of a fire. Nevertheless, other sources, both statutory and industry guidance, do clearly indicate the need for adequate levels of lighting.

Table 8.1   *Selected standards and guidance*

Standard or guide	Comment
Disability Discrimination Act (DDA)	This act requires reasonable measures to be taken by employers, and those in charge of buildings, to avoid those with an impairment from being disadvantaged in places of public access.
The Workplace Health, Safety and Welfare Regulations 1992	Regulation 8 states that "suitable and sufficient" lighting by natural light should be provided in workplaces (so far as reasonably practicable). The Regulations comment on excessive light and glare, the need to light stairs, and the arrangement of artificial lighting so as to avoid discomfort
Institution guides	CIBSE *Code for lighting* (CIBSE, 2002) and related lighting publications
HSE guidance	HSG38 *Lighting at work* (HSE, 1998)

Some associated lighting issues are relevant to the provision of a safe walking surface, particularly on stairs; see Table 8.2.

Table 8.2   *Design issues*

Design issue	Comment
Level of lighting	Natural as well as artificial Background, and task-related lighting
Variation in lighting	Uneven levels of lighting (eg a dimly lit area adjacent to one with a brighter level of lighting) where the contrast can contribute to unsafe environments
Means of locating and switching on lights	Regard needs to be had to the Disability Discrimination Act (DDA), and the elderly in particular Switches may need to be duplicated, eg top and bottom of stairs, or at each end of corridors
Night lights	Provision for lighting at night to assist the elderly
Glare	Design to avoid glare or dark areas Reflections off shiny floors can be avoided by careful design Effect of lampshades in reducing illumination

An overview of lighting in respect of the elderly (Easterbrook *et al*, 2003) concluded that:

- fluorescent lighting appears to be particularly well suited to older persons (noting that designs other than strip lighting are now available)

- visual clues (eg use of colour) can assist with walking

- higher levels of lighting are needed for older people, as well as in public places and in the home.

8.3   NOISE

Although there is a dearth of research data determining the relationship of noise to pedestrian slip accidents, there is some information (Cox and O'Sullivan, 1995) to suggest that excessive ambient noise may be a contributory factor in the causation of accidents.

There is also some evidence that a sudden, startling noise may temporarily impair performance. On the other hand, it is suggested that a varied acoustic environment supports a blind person's mobility.

Apart from the direct effect of noise, it also makes communication more difficult, perhaps masking sounds that are trying to give warning or direction. US research (referenced in Cox and O'Sullivan, 1995) concluded that, for speech to be heard at 6 m, the noise level in a room with no reverberation should not exceed 50 dB(A).

Noise in buildings is covered by the Control of Noise at Work Regulations 2005 and the Building Regulations. The former is only concerned with ill health in the workplace that arises as a consequence of noise. However, in specific circumstances the general workplace risk assessment, implemented as part of determining a safe method of work under the Management of Health and Safety at Work Regulations 1999, might identify "inability to hear instructions" as a risk. The Building Regulations are concerned with transmission of noise through floors and walls and not with guarding against levels of noise that might affect the likelihood of pedestrian slips.

8.4 VISUAL AND OTHER DISTRACTIONS

A pedestrian's attention to conditions can vary depending upon distractions such as obstructions, signs on or adjacent to the surrounding surface, and behaviour such as use of a mobile telephone or sharing a conversation. All these can increase the likelihood of a slip.

Vision is vital to normal walking and is especially important when other sensory input is reduced, as when elderly or impaired in some other way. Vision provides information about the movement of the head and body relative to the surroundings and is important for the automatic balance responses to reflect changes in surface conditions, such as a change to the alignment of the surface.

Figure 8.3 *Visual indication of a change in level (courtesy HSE)*

Those distractive aspects related to the design and management of the surface and its surrounds can be controlled, preferably by giving consideration at the design stage for a new surface. Those that relate to human behaviour can in some circumstances be controlled through guidance and training to the users, as indicated in Table 8.3

Table 8.3   *Control of visual distractions*

	Distraction	Comment
Design-related	Obstruction	Have regard to the placing of physical obstruction that might give rise to a slip, eg opening doors into a pedestrian pathway, work activities in close proximity to pedestrian usage
	Signage	Avoid the placement of signs that are unnecessary or difficult to read, thereby taking attention away from the act of walking
Management-related	Behaviour, eg use of mobile telephones, conversation, running	Employees and visitors should be warned if hazards exist in, for example, an industrial environment
		Aim should be to design-out in the first instance, as far as reasonably practicable

8.5 CONDENSATION

Condensation (which is a form of contamination, as noted in Chapter 5) can form on certain walking surfaces if the conditions are appropriate.

High internal relative humidity levels will result in condensation of water vapour on cooler vertical and horizontal surfaces. This condensate may run down the wall and onto the floor, derive from an overhead source (including ceilings), or form directly on the floor itself. The condensation may subsequently freeze.

Typical sources of condensation might include, for example:

* glazing
* pipework
* metal surfaces.

As with other forms of contamination, it should be anticipated and "designed out" or dealt with as part of the management regime, so far as reasonably practicable.

Human factors

> **KEY POINTS**
>
> ◆ *Managerial commitment to safety is a pivotal influence on safety culture.*
>
> ◆ *Risk management strategy should be implemented.*
>
> ◆ *Slip investigations should aim to identify the causes.*
>
> ◆ *Slip incidence can be reduced by modification of behaviour.*
>
> ◆ *With appropriate warning human behaviour will be modified.*
>
> ◆ *Learned irrelevance contributes to lack of required behaviour modification.*
>
> ◆ *Obvious hazards may be overlooked due to other distractions.*
>
> ◆ *Impairment of mobility through pushing/pulling/carrying contributes to a significant number of STFs in the UK workplace.*
>
>
>
> Courtesy HSE

This is an area where the HSE has identified a need for further research. Information in this chapter has been obtained from a variety of sources, from the UK and overseas.

9.1 INTRODUCTION

While some factors that affect the likelihood of a slip occurring are both tangible and quantifiable, it is known that the people who themselves suffer the incident are a complex part of the total system that initiates it. They should be considered as both physical and psychological elements of the system (Rossmore Group, 2003).

The Liberty Mutual Research Center for Safety and Health, set up 50 years ago in the USA, carries out investigations into four areas of occupational health and safety, of which "slips, trips and falls" (STFs) is one. An article published in 2002 (Maynard, 2002) similarly identifies the complex causes of STFs and looks briefly at the following.

Ergonomics – tasks performed have a direct effect upon the force characteristics associated with a person's behaviour, movement patterns and gait.

Biomechanics – the study of the mechanics of the body and the way it interfaces with a surface while walking is relevant to the dynamics of slipping.

Psychology – human perception and response to slippery conditions affects the likelihood of a slip occurring. If a slip hazard is recognised, generally gait will be adjusted accordingly. If the slip hazard is not perceived the failure to make this adjustment increases the likelihood of a slip occurring, possibly followed by a fall.

(An HSL report into tripping notes that forewarning of a trip hazard results in subjects significantly modifying their gait, with the toe clearance being increased by up to 50 per cent (Loo-Morrey and Jeffries, 2003). There is a strong likelihood that similar statistics would apply to slip hazards.)

Tribology – the study of the interaction of sliding surfaces, from the Greek *tribos* ("rubbing"). In the field of STFs it is relevant to the study of friction between the shoe and the floor, the wear of these surfaces over time and lubrication at the interface.

It is acknowledged that people with impairment are more likely to suffer from STFs. It is estimated that 15 per cent of the UK population suffer from some type of impairment, and this is not only within the ageing population (Rossmore Group, 2003). "Impairment" in the context of STFs includes not only disability but also able-bodied people who are impaired by carrying, pulling or pushing. The total percentage of the population affected is thus in excess of 15 per cent.

It can be seen from the above that when investigating a slip incident (generally slip-and-fall) it is important to avoid reaching a conclusion as to the cause without thoroughly investigating all conditions at the incident location at the time of the slip. It has been suggested that this should include "extensive" interviews with the "victim", together with measurements of the slip-resistance of the floor and the light intensity and contrast (Miller, 1998). Investigation of cleaning methods would also be needed here. The nature and condition of the footwear involved may also be a contributory factor in such incidents. It is thought that incident records do not include such information, certainly in rigorous detail. A sample of 100 entries from the London Underground Incident Capture and Analysis System (INCA) contained 15 slip incidents, of which two referred to a passenger "wearing high heeled shoes" and one to a "female...wearing leather flat rubber soled shoes". Behaviour such as horseplay, particularly when there has been significant alcohol consumption, may also be a contributory factor.

> A key recommendation proposed in the 2005 BBC Reith Lectures, given by Lord Broers, was that accident investigation should concentrate on finding the cause of the accidents not the person or persons to blame.

In the context of railways it is recognised that the transport of over-sized baggage is a contributory factor to STF incidents at stations. Dynamics are also important, as there are likely to be enhanced risk factors attributable to a combination of people hurrying in crowded locations, sometimes on wet or otherwise contaminated surfaces, often impaired by burdens and possibly uncertain as to where they are required to be and thus looking at (or for) signs at high level at the expense of recognising surface hazards. This has been verified during employee workshops (Rossmore Group, nd) conducted at York station (GNER) and Crewe station (Virgin), both of which are busy interchange stations.

Figure 9.1   *A busy station platform (courtesy Deborah Lazarus)*

Hospitals (and other healthcare premises) may also be characterised as locations where there are enhanced risk factors. These may include people with a range of disabilities and infirmities, often hurrying to an appointment in unfamiliar surroundings, where signage may be a distraction from any hazards at floor level.

Figure 9.2   *Hospital entrance – signs at different levels (courtesy HSE)*

It is recognised that further research is needed in this area. Some is being conducted explicitly in the context of railway premises. Rossmore Group is currently researching behavioural issues associated with STFs, described later in this chapter, and studies have been carried out in relation to certain population groups. Where possible, existing work has been drawn upon in the following sub-sections.

Figure 9.3   *Railway stations present a variety of slip hazards (courtesy HSE)*

9.2   ELDERLY

The terms "elderly" and "older people" do not cover a single homogeneous group. There are large variations in age, occupation, health and mobility. With rising life expectancy the group is going to grow, however defined, and it is important that its members should be able to lead healthy and independent lives for as long as possible.

Statistics (from various sources, within which "elderly" is not defined) show that slips, trips and falls occur more commonly among the elderly. This is because reaction time increases with age, the time needed for recovery from a heel slip is longer, and because muscle strength is reduced. The first of these makes slip more likely to result in a fall, which in turn may have more serious consequences for the elderly because their bones are more brittle. The second characteristic means that recovery from the initial slip is harder, also tending to initiate a fall.

The increase in STFs is much more noticeable in older women. While this has been ascribed in the past to post-menopausal changes, it is now considered that these are insufficient to account for the increase in numbers. In research targeted at trips it is suggested that it is "a combination of physiological and 'life-style' factors", which may be used to explain the increase in trip incidents in older women. Only limited information is available in this area, however (Loo-Morrey and Jeffries, 2003). While it may be assumed that similar issues apply to slips, there are perhaps even fewer data.

STF incidents that end up being reported in the railway SMIS (Safety Management Information System) are probably only the tip of the iceberg, with many incidents going unreported because of the attitude of the victim. There are many eye witness accounts of falls incidents where the victim is able to recover from the fall with minor injury or damage to clothing and proceed on their way. These normally go unreported because the victim may feel stupid or be hurrying to make a connection – probably one of the reasons for the fall in the first instance. It is commonly noted that younger, fitter individuals will behave in this manner. The occurrences of STFs in the elderly, however, tend to be reported since the outcome tends to be generally more serious. There is also an issue here in reporting systems – staff are normally required to report an accident they have seen only if it has resulted in first aid or medical treatment being administered (Rossmore Group, nd).

The DTI has issued guidance entitled *Avoiding slips, trips and broken hips* for both older people and the professionals working with them (DTI, 1999; DTI, 2000). The Department has also funded research on how older people use their stairs, conducted against a background of 57 000 older people falling on steps and stairs in the home each year. Broken bones in older people caused by a minor bump or fall (fragility fractures) are likely to result from osteoporosis. Ninety per cent of people who break their hip do so following a fall. Data from 1990 showed that hip fractures cost the NHS an estimated £160 million annually.

A laboratory study reported in the *International Journal of Industrial Ergonomics* investigated age-related effects of both obstructions to view and transitions in floor surfaces on gait characteristics related to slips and falls (Bunterngchit et al, 2000). The study was carried out on a small population – 10 each of a younger group (college students) and an older group (over-65s, all "in good physical shape"). The study recognised that this population could not be taken as fully representative, and that some of the changes identified were small. It was, however, felt that the study identified issues that should be recognised more widely and highlighted some areas where further study was needed:

- carrying even a light load may increase the risk of slipping

- while certain types of floor surface are recognised as being more slippery, there are also implications for walking on transitional floor surfaces

- walking on transitional surfaces and carrying light loads changed the gait patterns of elderly individuals more than the younger population

- further study is needed to see if the gait characteristics of less active elderly people are changed significantly by transitional floor surfaces and/or carrying objects that obscure the visual field at foot level.

9.3 PUSHING, PULLING AND CARRYING

Much of the existing research on pedestrian slip risk assessment is limited to the study of able-bodied pedestrians moving on level ground. It was considered that a significant proportion of the STF injuries incurred in the UK workplace would involve pedestrians carrying loads while moving.

Figure 9.4   *Pedestrians with a variety of encumbrances*

Earlier research has shown that the dynamics by which heel impact occurs appear to be responsible for whether or not a slip happens. It was postulated that modifying the impact dynamics used during walking by the presence of a encumbrance would change the probability of slip occurrence.

The HSL has carried out research into the effects of central and off-centre encumbrance on pedestrian slipping. Laboratory-based investigations into this aspect of the pedestrian slipping problem suggested that:

- encumbrance increases the risk of pedestrian slipping

- off-centre encumbrance (eg a bag carried via a shoulder strap) poses a higher slip risk than central encumbrance (eg a centrally positioned rucksack) (HSL, 2002).

The results did not show good correlation overall for varying floor and shoe combinations. The research report suggested that further work should be carried out using a wider constituency of varying mass, gender and height, or for one individual looking at a greater number of variables with respect to surface inclination, shoe sole and flooring materials. This is being progressed through a joint BRE/Rapra project.

External influences may also be significant when pedestrians are already encumbered. These include signage outside the normal field of vision, bright lights and sudden noise, any of which may distract and could lead to loss of balance, or prevent a hazard such as a change of level or surface registering.

9.4 PEDESTRIAN GAIT

There are gait patterns that make walking on slippery surfaces less hazardous, assuming that the slippery condition has been identified before reaching the surface. Clearly running, jumping and sliding should all be avoided. The recommended stance is to take slow, short steps with the toes pointed slightly outward. For additional balance, it is important to keep the hands free – unencumbered and out of pockets – and at the sides for support if a fall is initiated (Lehtola *et al*, 2001).

A study in Hong Kong investigated gait pattern for subjects walking on potentially slippery surfaces (Fong *et al*, 2004). This was focused on construction site conditions, against a background of 30.1 per cent of traumatic injury occurrences, the highest in sector ranking, being recorded on construction sites.

An initial survey was conducted so as to identify the surfaces, contaminants and footwear that best simulated local site environments. Sixteen conditions were identified:

- two types of surface – wood and cement

- four types of contaminant – dry (not defined), sand, water and oil

- footwear – safety shoe and lightweight cloth sports shoe.

Twelve males, all right-legged and free from both injury pain and abnormal gait, participated in the walking test, which was preceded by a mechanical slip resistance test to determine the dynamic coefficient of friction (DCOF) of the surface conditions. The test used a pulley system to drag a weighted shoe on a test area of walkway surface over a force plate. Weights were added gradually until the shoe started to slide. Recording of the corresponding horizontal and vertical loads from the force plate in turn allowed calculation of the DCOF.

The gait pattern parameters investigated were:

- stance time
- swing time
- stride time
- stride length
- heel velocity at contact
- mean propagation speed.

Tests where slips occurred were discarded for this particular study: the intention was to identify changes in gait pattern as adjustment to slippery conditions took place. The walkway conditions were divided into three groups, classified as "resistant", "unsure" and "slippery" using the DCOF results.

The study showed that the first three of the above parameters had significant correlation with the DCOF of the walkway conditions, with increases in all three as the slip potential increased. Heel velocity at contact decreased as slip potential increased. Stride length and mean propagation speed tended to decrease as the slip potential increased from "resistant" or "unsure", and decreased again with a further increase in slip potential.

The findings identified changes in gait pattern as humans adapted to slippery conditions comparable with those found on construction sites.

Faster walking can be a major problem, as can running. In these motions there is a harder landing on the heel of the front foot and an associated harder push-off from the sole of the rear foot. These require a higher coefficient of friction to prevent slips occurring. Sudden changes of direction have a similar impact (Lehtola et al, 2001).

9.4.1 Walking

Walking, in common with most other motions of the body, involves its centre of gravity. This may be thought of as the "balance point" around which a movement operates, and has been estimated to be at a point about 55 per cent of an individual's height above the floor for an average individual when standing (Miller, 1998).

The location affects how an individual moves – including both walking and falling – and will change during various activities. During walking it lies alternately over the left and right foot. The normal walking pattern for humans is termed "striding bipedalism", because when standing and walking both feet are in contact with the walking surface (Miller, 1998).

There are more detailed descriptions of the mechanics of walking that are not included in this guidance. There are also numerous variations in walking pattern. The significant issue is to recognise that the motion of walking results in "successive losses of balance", with each incident being of short duration and offering opportunity for both recovery under normal circumstances or more significant loss of balance under certain circumstances, which may lead to a fall.

The centre of gravity must remain within a base area, which itself changes during motion, in order to avoid falling. If a slip occurs, the foot may be mispositioned such that the CoG moves outside the base area. This will generally be followed by an instinctive attempt to regain equilibrium by adjusting the body components, but a fall will occur if this is not successful.

9.4.2 Running

When a heel slip occurs, recovery is necessary to prevent a subsequent fall. This will be harder when running, because of the increased momentum and reduced reaction time compared with normal walking. Running also results in changes to stride length and heel contact angle in comparison with normal walking motion.

> **Sports star fell in hotel**
>
> If you think falls only happen to the elderly, think again. Even athletes in the best of health can fall victim. Former NBA All Star Robert Reid tells *Inside Edition* that in **14** years of playing professional ball he never slipped and fell while running up and down the court. But when Reid came off an elevator in a Houston hotel, he ran into a wet spot and fell. "My right foot hit that wet spot... and next thing you know I'm flying up in the air... and I came down and I blacked out." Reid, who needed two surgeries to repair his knee, sued the hotel and was awarded $285 000.
>
> Source: NFSI, 2002

9.4.3   *Slip and fall*

Slips are primarily caused by a slippery surface and compounded by wearing the wrong footwear. In normal walking, two types of slips occur. The first happens when the heel of the forward foot contacts the walking surface. Then, the front foot slips forward, and the person falls backward.

The second type of fall occurs when the rear foot slips backward. The force to move forward is on the sole of the rear foot. As the rear heal is lifted and the force moves forward to the front of the sole, the foot slips back and the person falls.

The force that allows a person to walk without slipping is commonly referred to as "traction". Common experience shows that dry concrete pavements have good traction, while icy surfaces or freshly waxed floors can have low traction. Technically, traction is measured as the coefficient of friction (CoF). A higher CoF means more friction and therefore more traction. The CoF depends on the quality of both the walking surface and the soles of the shoes (Lehtola *et al*, 2001).

9.5   COGNITIVE SENSES

Walking is a motor skill. Once learned, motor acts like walking are initiated in the cerebral cortex without conscious intention or intervention. A person does not need to tell their body what joint to move or which muscles to energise. The mind orders a whole action, and the details occur without conscious direction.

A national champion golfer was asked to describe the steps involved in making a shot. She said, "I see the shot, then feel it, and then I do it". Walking is the same.

Ordinarily, people do not consciously recognise changes in the walking surface. While walking, the pedestrian scans about 10–20 feet ahead of them. These observations operate below the level of the conscious mind.

Most slip-and-fall accidents are caused by unexpected changes in the walking surface. To become aware of the hazard, the change must impinge upon the pedestrian's consciousness in some way.

If a potential hazard is seen, it can usually be avoided. It is possible to walk safely on an icy pavement, for example, if the hazard is observed and the pedestrian alters their behaviour accordingly. It is the unseen hazard that generates maximum risk (Miller, 1998).

This demonstrates the importance of making people aware of such changes. Research into the possible methods of conveying this information and their relative efficacy would be of value. Research from the US notes evidence that humorous signs may be more effective than the simpler alternative. One example given is that "CAUTION – WET FLOOR" is less effective than "WET FLOOR – SKATE, DON'T SLIP" (Lehtola *et al*, 2001). It is not known how widely this has been tested.

Recent work, reported at the nineteenth Annual International Occupational Ergonomics and Safety Conference, suggested that signs that advise what to do to avoid the hazard are most effective (ISOES, 2005).

Figure 9.5   *Warning sign outside Beaumaris Castle*

From Rossmore's experience (observation and employee workshops in the railway station environment) it would appear that certain hazards such as a damp terrazzo surface may not increase the individual perception of risk, whereas contaminants such as bird droppings and other animal excreta will be avoided, although perhaps for reasons of hygiene rather than of increase of personal risk.

Slips may also be caused by hazards that are perfectly obvious but are not observed because of distractions, including:

- wearing sunglasses in dimly lit areas
- moving from bright sunlight into lower levels of illumination (or vice versa)
- looking for signs or other indicators above floor level
- using a mobile telephone
- carrying or pushing something that obstructs the view of the floor.

9.6   PERCEPTION

Much of the earlier research into slip resistance was carried out in laboratories. In 1994 the HSE initiated work on floors in use in an attempt to validate slip-resistance measuring instruments and to gain information on surface roughness of floors "in the real world" (HSL, 1994).

It became apparent from site visits that accident statistics were of limited use in assessing the performance of the floor as the reporting of incidents tended to be unreliable and incomplete. It was decided instead to obtain more subjective opinions from users.

A questionnaire was sent to each site visited asking floor users to rank the slipperiness of the floor. The results were compared with instrument readings obtained for the same sites.

The study showed that, in general, the perception of the user is influenced by:

- the type of shoe being worn
- the user's age
- the condition of shoes
- contamination on the floor
- the speed of motion across the floor
- the level of lighting.

The study also showed that floor users perceived the increased slipperiness of a wet floor.

At a retail bakery the staff perceived the floor to be dangerous while customers regarded it as perfectly safe. This is a consequence of the different gait of the user groups: the staff move quickly to serve customers who tend to move much more slowly.

9.7 ADDITIONAL BEHAVIOURAL RESEARCH

In April 2005 the HSE commissioned research into the behavioural aspects of STFs (Rossmore Group, 2005). This included a full literature search and review, and development of a hybrid framework or model to enable relevant research to be woven together. It is acknowledged that the specific research into behaviour associated with STF accidents is still quite limited and is probably little more than is contained within Rossmore's summary document. The Rossmore work is continuing to examine the elements of the developed model to identify other research that may have been conducted within other discipline areas and to identify its validity when applied to STF scenarios. This research will concentrate on behaviour of different stakeholder groups and be based in applied psychology. The initial research was completed in late June 2005 and concluded that, without the correct organisational climate, STF accidents will not be prevented through improvements in environment and equipment alone.

9.8 OTHERS

Disability has been referred to briefly in Section 9.1 above, and is also looked at within Appendix 2, which considers the legislative framework. Within the context of controlling slip risk the breadth of possible disability has to be considered. This will include those with impairment to their mobility, those who are partially sighted or blind, those with hearing impairment and also those with reading and behavioural difficulties.

In recognising this, due attention needs to be paid to:

- lighting
- sound levels
- appropriate surfacing (see also Chapter 10)
- legibility and location of warning signs
- clarity of the warning message
- issues such as adequate provision of grab rails.

In addition, individuals with some types of physical disability have higher friction requirements than the general population. When specifying flooring for areas where these individuals may form a significant proportion of the users, these additional friction requirements should be taken into consideration.

9.9 ORGANISATIONAL

9.9.1 General

It is recognised that policies and practices within an organisation can contribute significantly to reduce the number of accidents and injuries due to slips, trips and falls. The National Ag Safety Database (Lehtola *et al*, 2001) contains the recommendations set out in the box below.

Advice to staff

- Owners, managers and supervisors should commit to preventing slips, trips and falls

- regular frequent inspections of working and walking areas should be conducted to identify environmental and equipment hazards that could cause STFs. Special attention should be given to the working and walking surfaces, housekeeping, lighting, vision, stairways and ladders. Immediate corrective action should be taken

- extensive safety training on the prevention of STFs should be provided for all new employees. Regular retraining should be provided for all employees. Special attention should be given to proper walking, carrying, climbing and descending stairways, ladders, vehicles and equipment. Unsafe practices should be corrected immediately.

- all workers should wear proper footwear for their work and environment whether in the office, shop, plant, feedlot or field

- all STFs, with or without injury, should be reported, recorded and thoroughly investigated. Corrective action to prevent such a repeat occurrence should be taken immediately.

Source: Lehtola *et al*, 2001

The National Ag Safety Database notes that "Slips, trips and falls whether on or off the job are expensive, disruptive, painful, and may be tragic" (Lehtola *et al*, 2001).

The legal position of employers and other duty-holders in this area needs to be understood. There is case law that demonstrates that it is not an absolute duty to have a "safe" floor. Nevertheless, adequate systems and procedures need to be documented and implemented to demonstrate that reasonable precautions have been taken and to safeguard the position of the duty-holder.

The case quoted below, for example, illustrates that in circumstances where spillages seldom occur or are not foreseeable and where procedures exist for inspection, the occasional spillage will not make the employer liable.

Kim Lesley Furness v Midland Bank plc (2000) CA

(Simon Brown LJ, Sir Christopher Slade) 10/11/2000

It was not reasonably practicable for the defendant employer to instruct its staff to keep a look out for spillages on stairs where there was no history of spillages and the spillage complained of was a very small amount.

Appeal by the claimant ("F") against an order dismissing her claim for damages for an injury sustained in an accident on the defendant's premises. F was employed as a lending officer by the defendant. In 1997, shortly before F was due to finish work for the day, she slipped on water that had been spilled on stairs in the offices of the defendant, causing her injuries. F alleged breach of statutory duty by the defendant under the Workplace (Health, Safety and Welfare) Regulations 1992. At the hearing of the claim, it was found that F had slipped on a few small droplets of water which could have been wiped away with a single piece of tissue. It was further found that there was no history of leakage or spillage on the stairs where the accident occurred. The Recorder had concluded, making an order dismissing F's claim, that the degree of risk from such a spillage was extremely small, and to protect F from a fall would have required continuous supervision of the staircase, which, the Recorder found, was not reasonably practicable. F sought to show that the defendant had failed to take reasonable precautions against spillages on the stairs. What the defendant should have done, F argued, was instruct its workforce, or some of them, to keep a look out for spillages. In failing to do so, F claimed, the defendant had failed to discharge its statutory duty under the 1992 Regulations. The defendant argued that the appeal should be dismissed because it carried out health and safety checks periodically, and that the staircase was cleaned at the end of each working day.

Held: (1) The spillage of water on the stairs was clearly a substance which was capable of causing a slip or fall. The burden was therefore upon the defendant to establish that it was not reasonably practicable to keep the stairs clear of spillage (Nimmo v Alexander Cowan & Sons Ltd (1968) AC 107). (2) There was no doubt that it was reasonably practicable for the defendant to have instructed its employees in the manner suggested by the appellant. However, a failure to make such an instruction was not a breach of the 1992 Regulations. (3) If there were frequent spillages, it would have been necessary to have instructed staff to be alert and deal with spillages (Ward v Tesco Stores Ltd (1976) 1 WLR 801). (4) In cases such as the present one however, where spillage was very rare and the premises were used by employees only, it was absurd to consider that the suggested instruction served a useful purpose. (5) There was no requirement for the defendant to instruct its staff to keep a look out for spillages. It was not reasonably practicable for the defendant to keep the staircase free from spillages of such a small amount. (6) Accordingly, the Recorder reached the correct conclusion, namely that F had not established a breach of the 1992 Regulations.

Appeal dismissed.

Source: QBE Insurance

In contrast, the following case is one where the employer was found to be in breach of the Workplace (Health, Safety and Welfare) Regulations 1992.

McGhee v Strathclyde Fire Brigade 2002 S.L.T. 680 (OH)

A fireman M sustained injuries after slipping on a floor at work. The tiles on which he slipped had been cleaned and polished 30–60 minutes earlier. M averred that S had been in breach of its statutory duties under the WHSWR 1992 Reg 12, Reg 12(1) and Reg 12(3); that the floor was not suitable due to wear and the tiles not being of non-slip material, and that that S had failed to keep it clear from a substance which might cause a person to slip.

Held: that Strathclyde was liable; the freshly polished floor represented a real though relatively low risk of slippage in breach of Reg 12(3) and that the worn state of the tiles were such as to expose persons to a risk to their safety and in breach of Reg 12(1).

Source: QBE Insurance

9.9.2 Offshore industry

A research project was carried out for the HSE to develop a tool for use offshore by their field inspectors to consider factors likely to increase the incidents of STFs (Mason, 2003). The background to this was the rate of STFs in the offshore industry, identified as 26 per cent of all injuries offshore reported under RIDDOR between 1998 and 1999, of which 37 per cent resulted in major injuries.

It is considered that the principle of the tool, while developed for a particular need, may be applied more universally. It looks not only at the physical factors such as slippery or uneven surfaces, but also environmental aspects, the nature of the work being undertaken and, significantly, the organisational and cultural aspects.

The organisational issues included:

- clear roles and responsibilities
- housekeeping
- standard of supervision
- pressure to work fast
- inexperienced workforce
- poor perceptions of risks
- systems for reporting problems
- health and safety tour/agenda
- permit-to-work checks and sign-off following completion of work.

The project went through several stages of development, with trial and feedback being used to modify the initial checklist and user instructions. It was considered that the final tool could be developed into an HSE publication for general industry application, but at the time of going to press this had not been taken further.

Other related reports include the *Behaviour modification to improve safety: literature review* prepared by The Keil Centre (Keil Centre, 2000).

9.9.3 Construction industry

Many of the issues in Section 9.9.2 are of concern to other sectors and industries, particularly the construction industry (see Loughborough University and UMIST, 2003).

This is replicated in Hong Kong, where traumatic injury occurrences on construction sites are higher than in other industries. A study carried out to investigate slips on site is described above (Fong *et al*, 2004).

Guidance to prevent slips, trips and falls in the UK cement manufacturing industry was produced in June 2005 (Gibson, 2005). While this is targeted primarily at the industry's own employees, some of the recommendations and guidance are applicable more generally to contractors.

9.9.4 Food retail industry

The Health and Safety Laboratory has carried out further work related to the food retail sector. A scoping study (HSL, 2003) concluded that it was possible both to produce two discrete types of measure of safety and risk management performance, and to identify contexts where these might be used in complementary fashion.

The food retail sector shares with other sectors the need to develop a positive safety culture and institute appropriate safety management strategies to control slip (and trip) hazards. As with other policies, the issue of culture is important if the safety policy is to be successful. It is believed that the majority of staff will conform to workplace norms, and that these norms can themselves be shaped in particular workplace contexts.

The HSL study identifies some accepted definitions of safety culture, of which the following is proposed as being of relevance to this guide:

> ...the set of beliefs, norms, attitudes, roles and social and technical practices that are concerned with minimising the exposure of employees, managers, customers and members of the public to conditions considered dangerous or injurious.

(Turner et al, 1989)

There are several variables that have been found to impact upon safety culture, but it appears that there is some consensus over what is considered to be a core set, of which managerial commitment to safety is regarded as a pivotal influence.

Individual attitude assessment techniques were used to measure the performance of organisations and groupings of organisations.

The topics identified for inclusion in the survey were:

- profile of health and safety
- rules and procedures
- management commitment
- supervisors' commitment
- locus of control
- staff training and involvement.

A safety culture: hazards should not be ignored

Here an employee works around a hazard: effective reporting and remediation schemes should be in place that encourage the reporting of hazards.

Source: British Cement Association

Various tools were developed in the study:

- checklist for managers and supervisors to use in individual retail outlets as a guide to the safety of the store both overall and in different areas
- workforce attitude questionnaire
- safety audit checklist and information sheet
- proforma for safety improvement notes.

The study concluded that the developed behavioural safety audit measures can provide an active indication of shop-floor safety performance and can detect differences between:

- geographical locations
- types of retail activity
- store populations.

They are "considered to offer significant potential in support of a range of interventions and initiatives aimed at monitoring and improving slip and trip risk management within the food sector".

Inexperience in the workforce affects the likelihood of a slip-and-fall, as does age. The National Floor Safety Institute based in Texas produced the Restaurant Slip-and-Fall Accident Prevention Programme as part of its best practices project in 2003 (NFSI, 2003). Its work again shows that, while wet or otherwise "dangerous" floors are responsible for half the slips and falls occurring in the industry, human factors, including insufficient training, contribute to the remainder.

Part of the contribution comes from what is termed "inadequate hazard identification". It is strongly recommended that warning signs are used when surfaces are wet or otherwise contaminated, whether as a result of cleaning in progress, spillage or other causes. Evidence indicates that, with adequate warning, people will modify their gait accordingly. The yellow floor signs commonly employed are familiar in supermarkets, railway stations and cloakrooms, among others, but they are over-used, which tends to mean they are ignored. Being left in position after the floor has dried this does not itself pose a risk, although it is known that this contributes to the perception that they can be disregarded, but when the surface is wet there is an obvious problem.

Thermochromic warning sign

Although the following relates to a warning system for icy conditions, strictly outside the scope of the guidance, it is included because the signage provides a good example of giving a clear, unambiguous message.

The safety sign is designed for use in public places and provides a warning when temperatures are low. It is based on ink and printing technology. The main body of the sign is photoluminscent and is able to glow in the dark for 14 hours. The warning element is thermochromic, which means the ink starts to show when the temperature is below 5°C. There is a built-in margin for error to allow for the signs being placed on occasion in an area that may not be the coldest on site.

Courtesy ThermoSign, <www.thermosign.com>

Figure 9.6   *Warning of spillage: sign in place and cleaner approaching (courtesy HSE)*

Figure 9.7   *The sign very rarely moves away (courtesy Deborah Lazarus)*

Unnecessary wet-floor warnings

The study revealed that 65 per cent of the time you see a wet floor sign, the floor is not wet. Many business owners admit to such practice with the belief that such use assures them of a better defense in the event of a slip-and-fall lawsuit. Untrue. While the law requires that a property owner posts a warning in the event of a known hazard, the "failure to warn" rule also states that posting a warning when no such hazard exists is, itself, a bad practice.

Source: NFSI, 2003

The above box discusses an example of "learned irrelevance". There are many instances where learned irrelevance contributes to the lack of required behaviour modification. Other examples include false intruder alarms, false evacuation alarms and unnecessary motorway speed restrictions.

Training

Problems can be caused by lack of awareness among staff, carelessness, not understanding how slips or trips occur, lack of training etc.

It is important for companies to train, inform and supervise staff on points such as the significance of spillages, "cleaning as you go", reporting equipment defects, how to use and care for safety measures (including footwear), the importance of thorough cleaning and drying of floors, and reporting incidents as soon as they occur.

Advice to employees

It is important that all employees should be made aware not only of how to prevent these slip and trip incidents but also their own responsibilities in relation to "good housekeeping" and who to notify in the event of problems.

In order to maintain safe floor surfaces and walkways throughout the workplace.

- Keep all corridors, passageways, storerooms, and service areas clear of debris, boxes and storage. Never block these areas, even temporarily. Emergencies don't usually come with advanced warning and are not likely to give you time to clear cluttered escape paths.

- Keep stairwells clear at all times. Do not store boxes, files, or other debris in the stairwells or landings.

- Pick up dropped pencils, paper clips, and rubber bands that can cause you or a co-worker to slip.

- Contact building management if you see areas that are cluttered with rubbish.

- Wipe up spills immediately. If a spill is too large to clean up quickly, contact building management.

- Report uneven, defective flooring, worn spots in carpets, chipped tiles, and worn stair treads to building management.

- In areas where wet or damp conditions are likely to routinely exist, appropriate drainage should be maintained and grating, mats, raised platforms, or anti-slip strips should be evaluated and considered for control or prevention of slippery conditions.

- Every floor, work area, and passageway should be kept clear of obstructions that protrude into the walkway or have the potential to result in unsure footing, such as loose parts, boxes, packing material, or tools. This includes areas where construction or demolition debris has the potential of negatively impacting permanent or temporary walkways.

Getting everyone involved

- Make sure that everyone working in an area has a good understanding of the right way to work and the precautions to take.

- Regular supervision is needed to make sure staff are following instructions about safe practice.

- It is a good idea to involve employees at all levels when looking at risks and agreeing the safety measures needed. With this approach, you are more likely to successfully tackle the hazard and get employees to comply with any necessary action.

- Trade union safety representatives and representatives of employee safety should be consulted as they may identify problems or come up with solutions that you may not have considered.

Source: AT&T, 2005

Company commitment to health and safety

PAYE Stonework and Restoration believes that all members of the company from the top downwards should feel committed to good practice in relation to health and safety issues. As part of this it believes in treating everyone as part of the team, encouraging them to take care of themselves and also take responsibility for others around them, for the benefit of the team collectively. PAYE has supplied all its employees with a rucksack containing both a full set of PPE appropriate to the work of the company and also a concise, pocket-sized safety handbook, which it asked one of its operatives to illustrate, covering the various aspects of their work.

Source: PAYE Stonework and Restoration

Workers and management unite to tackle slipping problems

Nestlé UK employs around 2000 staff on its confectionery production site in York. In 1996 the reported percentage of slip and trip accidents was 33 per cent. A joint management/union initiative was started that year to reduce this unacceptably high figure.

Principles were agreed between senior company and GMB Union officials, and a period of six months was allocated to plan the campaign. Current HSE material was used for briefing, with all managers, team leaders and safety representatives participating in the process.

Once these briefings were completed, all employees were informed of the initiative.

The following were adopted as part of the process:

- identify the causes of slips and review specific solutions
- use posters, with individuals on each site being given responsibility for changing these regularly
- keep stocks of incident report forms by the posters
- institute reporting from each site safety committee on slips and trips.

In addition:

- all accidents were jointly investigated and logged on computer system, with a colour system to confirm close-out
- periodic reviews were held with GMB officials outside the company.

The impact was dramatic: although initially the number of incidents reported increased, as a result of the awareness campaign, the reversal was rapid, with a 60 per cent reduction in slips and trips within three years.

Source: HSE, <www.hse.gov.uk/slips/experience/workers-unite.htm>

KEY POINTS

◆ *The vast majority of floor surfaces have adequate slip resistance in clean, dry, conditions, on a level surface.*

◆ *This is not the case if the floor is wet, dusty or contaminated with liquids.*

◆ *The pendulum test used according to the UKSRG guidelines accurately determines the slip resistance of the floor.*

◆ *Surface roughness, Rz, is a useful complementary measurement that can easily be used to monitor floors in service.*

◆ *A surface roughness, Rz, of 20 microns is recommended for surfaces foreseeably contaminated with water. Greater roughness is required to combat more viscous contaminants.*

◆ *The slip resistance of flooring changes over time. Wear can reduce its surface roughness to the extent that it becomes a significant slip hazard if wet or contaminated. While surface roughness measurement indicates the changes, only pendulum testing will measure the actual change.*

◆ *Unless a profiled floor affords physical interlocking between the profile and the footwear, it will only be as slip-resistant as the surface roughness on the top of the profile dictates.*

◆ *Sudden changes in the slip resistance of a floor can be a slip hazard; if unavoidable, it should be denoted by a colour change in the surface.*

◆ *It is not possible to give a generic slip resistance value for a flooring material because its performance is dependent on its surface finish. Indicative values are included in Table 10.1.*

Courtesy Forbo Flooring

10.1 INTRODUCTION

This chapter presents the most common floor finish materials for internal and external pedestrian use. It also addresses the performance of processes used to improve the slip resistance of surfaces, such as mechanical roughening, chemical treatments and applied finishes.

For all types of generic flooring, the pendulum tester gives an accurate indication of the slip resistance of a floor surface. A floor's surface roughness is a supporting parameter in assessing slip resistance. Most standard floor materials have a roughness (or micro-roughness) of between 1 and 100 microns.

Research has also shown that hard floors require higher levels of surface roughness than soft floors in order to produce satisfactory levels of slip resistance (Lemon and Rowland, 1997), although the magnitude of this difference does not appear to have been quantified. This may be because soft floors give slightly under the weight of the walker.

It is not possible to give a slip resistance value or surface roughness for a particular material because the way in which the surface has been finished determines its slip performance. A generic floor material can have a wide range of performance.

10.2 FLOOR SUBSTRATE

This is the construction that supports the floor finish and so is common to all flooring types. The most common type of substrate in commercial and industrial buildings is concrete construction, whether a ground-bearing slab, composite construction or pre-cast concrete planks. Timber floor construction with timber joists and floorboards or timber boarding on top is more widely used in domestic upper-floor construction.

The substrate of a floor should be capable of withstanding all the structural, thermal and mechanical stresses and loads that will occur during service. Through the presence of appropriate waterproofing membranes it should protect the floor finish against rising damp. It should remain stable when protected by the floor finish and be provided with the necessary joints to enable it to do so. If the substrate does not remain stable the performance of the flooring will inevitably be affected.

As a general rule, a concrete substrate that is to receive toppings should be :

- free from laitance and contamination
- mature and dry
- finished with a strong and even surface, laid to falls if necessary
- laid to the accuracy required in BS 8204-1:2003
- structurally stable under operating conditions (Timperley, 2002).

BS 8204-1 gives recommendations for constituent materials, design, work on site, inspection and testing of concrete bases that are to receive *in-situ* wearing screeds of the following types:

- concrete (BS 8204-2)
- polymer modified cement (BS 8204-3)
- terrazzo (BS 8204-4)
- mastic asphalt (BS 8204-5)
- synthetic resin (BS 8204-6)
- magnesium oxychloride
- pumpable self-smoothing screeds (BS 8204-7).

It also makes recommendations for bases and levelling screeds that are to receive flexible floor coverings such as:

- textiles
- linoleum
- polyvinyl chloride
- rubber
- cork

and rigid floorings such as:

- wood block and strip
- laminate floor coverings
- ceramic tiles
- natural stone.

It applies to ground-supported and suspended concrete floor bases.

Hard and flexible floor coverings can also be applied over a wide variety of non-concrete substrates, including existing finishes. Guidance on the condition of these substrates is given in the standards contained in the appropriate sub-section of this chapter.

10.3 CHANGE IN SLIP RESISTANCE WITH TIME

The slip resistance of a flooring material can be expected to change in the course of its life. Some surfaces will differ from the factory condition as soon as they are installed, for example if the application of polishes is part of the installation process (Thorpe and Lemon, 2000). Flood grouting of ceramic floors may also lower slip resistance as particles of grout reduce the ex-factory roughness. Although most surfaces will wear smooth once in service, this is not the case for all. Some become rougher as wear causes the surface profile to develop. Consequently it is not possible to predict which in-service changes will occur and measurement is necessary. Regular surface roughness measurements are useful in monitoring such changes. The effect of any changes in surface roughness should be confirmed by pendulum testing.

Softer materials, such as lino and PVC, wear more quickly than hard ones. Materials that had adequate slip resistance initially may reach a stage where this is no longer the case, becoming slippery in wet or contaminated conditions. If wear results in the floor becoming rougher, for example if it contained an abrasive, its slip resistance would improve. The pattern of use of a floor will mean that the change in slip resistance is not uniform, but varies with the areas receiving the most use changing most quickly.

Figure 10.1   *Uneven wear on timber and metal stair treads and landings changes their slip resistance characteristics over time (courtesy HSE)*

For a material to maintain its slip resistance throughout its life it must retain its surface roughness. Some floors achieve this, such as those that contain hard particles throughout their thickness and are progressively exposed as the matrix wears away. If a floor has a superficial layer of grit to improve wear resistance, but the bulk of the material contains none, then the risk arises that the surface layer wears off in time, leaving a slippery surface underneath. This has been observed in some PVC slip-resistant systems (Lemon and Rowland, 1997), which can cause particularly dangerous conditions, as a slip-resistant floor is usually specified where slipping is a known hazard. If a slip-resistant floor loses its superficial properties, it could pose a greater danger than if a regular floor had been installed. For example, it was observed that the slip-resistant paint finish quickly wore off a steel floor grating and the carborundum grit became detached from a GRP grating (Lemon *et al*, nd). The slip resistance of both was significant reduced in consequence.

Figure 10.2   *Outdoor stairs – coated with epoxy/grit only in the centre and mostly worn off (courtesy HSE)*

The influence of the cleaning regime may cause a reduction in slip resistance with time, particularly where polishes and buffers are used. Repeated cleaning and buffing was found to cause a progressive reduction in slip resistance and surface roughness in PVC flooring (Lemon, Thorpe and Griffiths, 1999b). See Chapter 4 for more information on cleaning.

10.4   PROFILED SURFACES

Profiled surfaces are available in a wide range of materials, including metal, timber, rubber and PVC sheet, and ceramic tiles. They have a distinct surface texture such as ridges or blisters and are often used where there is perceived to be a risk of slipping, such as toilets, shower areas, swimming-pool surrounds and industrial premises.

While profiled surfaces afford slip resistance when they result in a physical interlock between the profile and the footwear, where no interlock occurs and it is possible for the footwear to skate across the profile, then it is the surface micro-roughness that determines the floor's slip resistance. The micro-roughness of the top of the profile is particularly important.

The surface profiles also give higher levels of solid-to-solid contact between shoe and floor in the presence of wet contamination, because of the ability of profiled surfaces to pool a volume of contaminant before complete submersion (see Figure 10.3). HSE's site experience indicates that significant levels of contamination may nevertheless remain on the upper surfaces of the profile for long periods after the pooled contaminant has been drained, and that even extremely small volumes of contaminant may lead to a sharp decrease in slip resistance. Research (Lemon, Thorpe and Griffiths, 1999a) showed that for profiled ceramic floor materials, the slip resistance did not appear to be dependent on the profile height, profile leading-edge radius or the waviness of the profile but on its micro-roughness. For improved slip resistance on these surfaces they must be rough on a micro scale, regardless of the profile. It is common for the highest points on a profiled surface to become worn with time, so they may actually be more slippery than a flat surface would have been.

Figure 10.3   *A typical raised profile flooring for contaminated areas (Lemon, Thorpe and Griffiths, 1999a)*

The preconception that a profiled floor is less slippery may not be correct and the efficacy of profiled floors cannot be assumed without further investigation and testing.

Experienced operators can evaluate profiled surfaces using the pendulum test. Roughness measurements from the tops of the profiles can also provide useful information. It may be necessary to move the pendulum in order to ensure that different initial slider impact positions are taken into account.

10.5   GENERIC FLOOR TYPES

The generic floor types have been split according to the listing in BS 8204-1. Some that were not included in this list have been added. Magnesium oxychloride flooring has been excluded because it is no longer widely used.

10.5.1   In-situ *cementitious wearing finishes*

Concrete has been the natural choice in a wide range of flooring applications but especially for heavy and medium use. Methods of finishing the surface of concrete slabs have changed significantly over the past 20 years, the principal one being the acceptance that an *in-situ* concrete floor can be finished satisfactorily using mechanical floating techniques that allow thin sheet and tile finishes to be applied directly to the concrete surface without the need for screeding. In buildings such as industrial workshops and warehouses where there may be substantially heavy traffic it is more practical to use high-quality concrete that is directly finished.

There are several techniques to finish the surface, including:

- power float
- hand or steel trowelling for small areas of flooring
- power-trowelling by machines fitted with steel blades to simulate hand-trowelling and produce a smooth, dense wearing surface (the most common concrete finish)
- power-grinding that reduces surface irregularities to produce a hard wearing surface suitable for medium-duty use
- tamped or brushed for a rough surface.

The slip resistance of smooth concrete surfaces is fair to good, depending on surface characteristics. Surfaces are slippery when wet unless a textured finish is applied or slip-resistant aggregate used.

SRVs are typically:

smooth finish	dry >75	wet 20–30
carborundum finish	dry >60	wet 30–50 (Pye and Harrison).

The surface characteristics of the concrete will change with time and use, especially if the floor is subjected to contaminants from industrial processes, as these may be absorbed into the surface of the concrete.

Wherever hard particles are added to cementitious floors it is important to ensure that the particles are evenly distributed and that there are no gaps, otherwise slipping could occur in such areas on an otherwise slip-resistant surface. A small area of 75 mm × 75 mm would be large enough to allow a slip.

10.5.2 Wearing screeds (high-strength concrete, granolithic)

High-strength concrete toppings comprise sand and cement with high-quality aggregate such as granite, hard limestone or basalt and are fully bonded to the concrete sub-base. Their composition and performance are covered in BS 8204-2.

They are used in commercial and industrial floors where high resistance to abrasion and mechanical wear is required. BS 8204-2 includes abrasion resistance categories. The highest abrasion resistance can be achieved on direct finished floors and wearing screeds by the skilful application of finishing techniques and the use of resin-based curing compounds, or in-surface sealers, or by the application of a thin dry-shake finish. The resistance to abrasion increases with the number of trowelling operations and the care with which they are carried out. However, excessive trowelling at this stage should be avoided to prevent the production of a polished or slippery floor. The additions of such dry-shake mixtures may decrease the surface roughness of the wearing surface of the concrete to such low levels (less than 10 microns) that it becomes unacceptably slippery when wet. If a suitable grit is incorporated in the dry-shake mixture, this may be avoided.

The flooring should be finished to produce a reasonable slip resistance for the expected use. Any of the following methods may be used, provided that the slip resistance value (SRV) of the floor surface is not less than 36 when tested by the method described in BS 7976-2:

- trowelling
- grinding the hardened surface to a fine-textured finish
- mechanically roughening the hardened surface, eg by shot-blasting or scabbling
- trowelling in, or incorporating in the concrete or screed material, slip-resistant granules
- providing slip-resistant inserts in the surface (for small areas only, eg ramps and stair tread nosings).

10.5.3 Polymer-modified concrete

Polymer-modified cementitious flooring is a mixture of cement, aggregate, water and polymer fully bonded to a concrete sub-base. Widely used polymers include SBR (styrene butadiene rubber), acrylates and vinyl acetates. Typically they are added to the flooring mix as an aqueous dispersion. For wearing screeds, the proportion of polymer solids should generally be within the range 7–15 per cent by mass of the dry weight of the cement. Their composition and performance are covered in BS 8204-3.

This type of flooring is used in industrial applications where higher moisture, chemical and impact resistance are required. Polymer-modified cementitious screeds are designed to be applied to bases of the following types:

- direct-finished concrete slabs
- fine concrete screeds
- existing concrete floors within buildings
- other types of bases for which there is a history of suitable use.

They can be used either as a levelling screed (to take another surface finish) or as a wearing screed (the surface finish). Where used as a wearing screed the following service conditions should be considered:

- the temperature to which the flooring will be exposed
- the nature and duration of any chemical exposure (contaminants, cleaning agents, oils, greases etc) likely to be in contact with the flooring
- wet or dry conditions
- slip resistance requirements
- ease of cleaning (including hygiene requirements).

Their slip resistance is reduced in wet conditions, but can be enhanced by the addition of applied surface additive (Taylor, 1998).

Figure 10.4 *Polymer-modified concrete flooring (courtesy Fosroc)*

10.5.4 Terrazzo and conglomerates

Terrazzo is made from pigmented white or grey Portland cement and marble aggregate and laid on a concrete sub-floor. It can be cast *in situ* or made into tiles on a concrete backing. Terrazzo tile and slab must comply with BS EN 13748-1. The cast-*in-situ* material is installed according to BS 8204-4 and the tile is installed according to BS 5385-5. It is honed, ground and polished to give a smooth finish, and is available in a wide variety of colours. It is typically cast in bays of 1 m², and is 12–15 mm thick, or in tile format usually 35 mm thick.

Conglomerates are manufactured from crushed rock concrete with cementitious or polymeric binder and cast into slabs to simulate a marble or granite finish. Unlike terrazzo, which is finished *in situ*, conglomerate is supplied as finished.

Both materials are used in a wide variety of applications, from domestic environments to operating theatres and highly trafficked public spaces, such as railway stations and shopping centres, on account of their good wear resistance and cleanability.

Figure 10.5 *Terrazzo flooring (courtesy HSL)*

Terrazzo generally has satisfactory slip resistance in the dry (BS EN 13748-1), but the combination of very smooth floor surface and hard smooth heel or sole material can be slippery even when dry. Similarly, dry contaminants such as dust, fibres, lint and paper can make the surface slippery (BS 8204-4). As the surface is typically polished it is usually very slippery when wet. The flooring should be finished to produce a reasonable slip resistance for the expected use. The SRV – defined in some standards as the PTV (pendulum test value) – of the floor surface should not be less than 36 when tested in accordance with the method described in BS 7976-2. To obtain the necessary degree of slip resistance in wet conditions it will be necessary to use a coarser final grind and to avoid a highly polished finish.

The choice of aggregate influences the slip resistance. Grit such as carborundum, bauxite silicon carbide or aluminium oxide may be added to improve slip resistance. The grit is harder than the main aggregate, so as the tile wears the grit wears more slowly and protrudes from the surface, giving an overall increase in roughness. Such surfaces may require different cleaning methods.

The normal finish of terrazzo obtained by grinding and polishing to a fine finish should be suitable for most dry, clean locations. On staircase treads and ramped surfaces, prefabricated anti-slip strips set in the terrazzo in a colour contrasting with the background should help to increase grip, so long as the strips are sufficiently close together.

10.5.5 Mastic asphalt

Mastic asphalt comprises a mixture of asphalt binder and graded aggregates. BS 8204-5 includes information on the types and grades of materials and the design of mastic asphalt flooring for use as an underlay or wearing screed on a concrete base or screed. It is applied hot and cools within two to three hours, providing a seamless finish. Traditionally available in black, dense reds and greens, it has been developed and is now specified in a series of vibrant colours, see Figure 10.6. The colour exists through the entire thickness and is not a surface application, so it will not wear through.

A well-laid mastic asphalt floor of the correct grade provides a hard, durable surface that is long-lasting, easily maintained and is not inherently slippery. It is able to carry heavy loads, provided that care is taken to avoid static point loading, which might cause indentation. It acts as an impermeable layer to liquid water and vapour. When mastic asphalt flooring is combined with a mastic asphalt waterproofing membrane, it can be used in wet process areas or in areas where washing down is a requirement.

There are four grades (I to IV) of flooring asphalt, which are selected according to the anticipated severity of use. Flooring grade asphalt should only be laid in heated buildings, because it is damaged by low temperatures. Paving grade is suitable for unheated or external conditions.

Mastic asphalt tends to have a rough surface with an open texture that softens with temperature (HSE, 2004b), which is inherently not slippery (BS 8204-5). Newly laid asphalt may have a smooth surface, while weathered asphalt tends to be matt. Slip resistance can be improved by initial sand rubbing during laying to prevent a smooth surface finish forming. The surface characteristics of asphaltic and other bituminous surfaces can be significantly altered by localised contamination, especially by petroleum products, to the detriment of any inherent slip-resistance properties. Frequent polishing will also reduce the slip resistance of the surface. Cleaning contractors may apply a lacquer to an asphalt floor, which is likely to be detrimental to its slip resistance.

Figure 10.6 *Mastic asphalt flooring (courtesy Pure Asphalt)*

10.5.6 Synthetic resin

BS 8204-6 describes the different types of resin-based floor finishes, from thin floor seals (type 1) to heavy-duty screed flooring (type 8). Most commonly the resin is epoxy, but polyurethane and acrylic are also used. They are applied to appropriately prepared concrete floors and may be from 150 microns to more than 6 mm thick. They may be applied by brush or trowel and some are flowable. The thinner paint-on finishes often require fairly frequent reapplication. As in all topically applied finishes, it is vital that the surface on to which the resin is to be applied is clean, dry and free from dusts.

Resin finishes are widely used in many industrial locations, such as within the chemical and petrochemical industry, the food, brewing and soft drink industries, storage facilities and warehouses, laboratories, workshops, hospitals, kitchens, cold stores and showrooms. They are chosen for their speed of installation, durability, impermeability, cleanability and robustness. In the food preparation industry resin floors have been successfully used where there is a need for seamless coatings that are impervious, aseptic, mould- and bacteria-resistant and easy to clean.

Wherever hard particles are added to resin floors it is important to ensure that the particles are evenly distributed and that there are no gaps, otherwise slipping could occur in such areas on an otherwise slip-resistant surface. An area of just 75 mm × 75 mm would be large enough to allow a slip.

Figure 10.7 *Painted concrete floor (courtesy HSE)*

BS 8204-6 states that the flooring should be finished in a manner that produces a slip resistance compatible with the circumstances of use. The SRV should be not less than 36 when tested, wet or dry as appropriate for the anticipated service conditions. While resin-based flooring can be formulated to produce smooth, non-porous surfaces with excellent slip resistance under dry conditions, the surface has to be micro-rough if it is to have adequate slip resistance under contaminated conditions. The heavier the likely build-up of contaminants, the coarser the surface texture has to be to retain the required level of slip resistance. Special consideration should be given to the slip resistance of ramps.

Surfaces with good slip resistance may be produced with a resin matrix and an aggregate – typically quartz, silicon carbide, carborundum or sharp sand – to give a very rough surface profile. The aggregate breaks through the contamination present to give shoe-to-floor contact. Aggregates may be up to several cubic millimetres in volume.

Such surfaces may be produced in three ways:

- the matrix is applied by brush or roller to the existing floor surface, and aggregate is then broadcast on to the matrix material
- the matrix and aggregate are mixed by the manufacturer, which is then applied to the floor as a single product
- the matrix and aggregate are applied on to a substrate such as sheet steel and sealed with a coating of matrix. It is marketed as retro-fit panels, which are securely fastened to an existing floor.

Specialist advice is necessary to quantify the type and quality of aggregate required to break through the anticipated contaminant. Specialist sub-contractors should always be used who are fully conversant with manufacturers' laying conditions. Whatever the aggregate size or type, the suitability of the cleaning regime should always be reassessed.

The HSE tested a range of these types of floor surface (Lemon, Thorpe and Griffiths, 1999b) using the DIN ramp test. It was not possible to measure some surface roughness because of the aggressive surface profile causes failure of the surface probe.

The following conclusions were drawn:

- the typical SRV for these materials was from 90 to more than 100 (Lemon, Thorpe and Griffiths, 1999a)

- the use of a larger grit size does not necessarily improve the slip resistance of these materials in wet conditions (this may not be the case for more viscous contaminants)

- if the manufacturer's installation instructions are not followed, the slip resistance of the surface may be reduced

- for materials where the aggregate was broadcast into the matrix, shedding of the aggregate occurred during testing, which resulted in significant reductions in their SRVs

- the use of coated panels produced very slip-resistant surfaces (CoF = 1.0), regardless of the aggregate size used

- reducing the number of grit particles per unit area reduces the slip resistance.

Poor application of anti-slip flooring

Following a number of serious slip incidents in the recently refurbished changing rooms of a leisure centre, the HSL undertook a series of *in-situ* pendulum tests. Although the floor was suitable for the application in terms of its SRV and surface micro-roughness, testing revealed that there were areas where the grit was not properly distributed in the resin, leading to very slippery bald patches.

Once the flooring contractor had reapplied the flooring in these areas, no more slip accidents occurred.

Source: HSE, <www.hse.gov.uk/slips/experience/swimmingpool.htm>, accessed 1 Dec 2005

10.5.7 Pumpable self-levelling/self-smoothing screeds

There are two types of self-smoothing screeds. Some have cementitious binders and others are based on calcium sulfate (gypsum). They are supplied as proprietary formulations, either as a ready-to-apply liquid or as a powder, to which water must be added.

They are widely used as levelling screeds, but can also provide the wearing surface itself in internal flooring. Cement-based pumpable self-smoothing screeds are often chosen when there is a requirement to lay a thin screed section, when the screed is to be a wearing screed or receive a synthetic resin flooring for an industrial floor, or when quick drying is essential. Calcium sulfate-based pumpable self-smoothing screeds tend to be chosen when the screed section is to be thicker, when there is a requirement for an unbonded or floating screed or when the programme allows a longer drying-out time before application of floor finishes. Calcium sulfate screeds are generally not suitable for use as wearing screeds, nor are they suitable for use in damp conditions. BS 8204-7 provides details on the choice of materials, design and installation.

When they are dry, pumpable self-smoothing screeds will have adequate slip resistance. If slip resistance is required it is usual to apply a floor finish to the screed. For determination of the slip resistance the method of test given in BS 7976-2 should be used and the SRV (or PTV) should be not less than 36 (wet or dry).

Figure 10.8 *Pumpable self-levelling wearing screed (courtesy Ronacrete)*

10.6 FLEXIBLE FLOOR COVERINGS

10.6.1 Carpet

Carpet comprises a backing that supports a fibrous front face. It may be supplied in tile or sheet form. The backing is typically bitumen, PVC or foamed rubber. The fibres may be wool and other natural fibres or synthetic, typically nylon or polyester, or a mixture of these. They may be attached to the backing in a variety of ways such as adhesion or fusion bonding, weaving or needle punching. Fixing methods vary, from the traditional underlay plus gripper rods for woven carpet, to adhesive bonding of the tile to the substrate.

There are many British and European standards that apply to the wide range of different carpet types. BS 5325 is a code of practice for the installation of textile floor coverings. It covers woven, non-woven, tufted, tiles and others and recommends the suitable application techniques including gripper rods, fully bonded, release systems, full adhesion with carpet underlay and perimeter fixing.

Carpet is widely used in domestic, retail and office locations.

The risk of slipping on properly installed carpet is low, provided it is well maintained. Carpets that contain a proportion of natural fibre have slightly better slip resistance than 100 per cent synthetic materials. However, the risk of slipping increases if the carpet is not cleaned to the point that it develops "smooth" areas where the pile is stuck together, or where it is worn smooth. In the extreme, wear also creates trip hazards.

The ease of use of carpeted surfaces for disabled users needs to be considered. Deep-pile carpet should not be used and the backing should not be too soft. Where wheelchairs might turn, and at other heavily used areas, the carpet should be adhesively bonded to the floor (BS 8300). Studies have shown that thick-pile carpet increases the likelihood of falls in older people, although it also reduces the severity of the injuries incurred. Further research was considered necessary to establish whether the benefit of falls of reduced severity outweighed the increased likelihood of falls occurring. Similar results were obtained in a study comparing the use of vinyl and looped nylon carpet in a rehabilitation ward for older people, whereby fewer falls occurred on the vinyl, but they were more severe (Easterbrook et al, 2001).

In public buildings, surface-laid rugs and mats should not be used. If in exceptional circumstances they are used, they should be securely fixed to the floor at their edges and at any joints, to avoid tripping or slipping (BS 8300). Carpet that is not well installed or is loose or worn is likely to increase the risk of trips and falls (Taylor, 1998).

The pendulum test can be used to measure the slip resistance of carpets, provided care is taken to set the footprint accurately. Given the fibrous nature of the carpet surface, it is not possible to use surface roughness measurements in an assessment of its slip-resistance (HSE, 2004b).

Figure 10.9   *Worn carpet with potential trip hazards (courtesy HSE)*

10.6.2   Linoleum

Linoleum is a natural flooring material made from oxidised linseed oil, pine rosin, wood flour, fillers and pigments. It is available in tile or sheet form with a jute hessian (sheet), polyester (tile), foam or cork backing (acoustic ranges). Linoleum is bonded to the sub-floor using either spirit-based or dispersion adhesives. European standards BS EN 548, BS EN 686, BS EN687 and BS EN 688 cover plain linoleum, foam-backed linoleum, cork-boned linoleum and cork linoleum, respectively. BS 8203 provides guidance on the installation of resilient floor coverings, including linoleum, on a wide variety of substrates.

Linoleum is classified under EN 685 as suitable for very heavy commercial use and general use in light industrial areas. It is frequently used in residential, healthcare, education, office and light industrial applications. The natural ingredients of linoleum give the product good anti-bacterial properties, so linoleum is commonly used in healthcare facilities, in conjunction with other good management techniques, to help to control the spread of hospital-acquired infections, including MRSA.

While it is not possible to incorporate a slip-resistant surface finish on to linoleum, the material does have relatively good slip resistance properties and, if properly managed by the user, performs well (Taylor, 1998). When tested using the pendulum method, linoleum gives a high dry slip resistance and a moderate to low wet slip resistance (test data should be obtained from the linoleum manufacturer.) According to BS 8203, it is also suitable for use on stairs, provided slip-resistant nosings are installed.

Figure 10.10 *Linoleum flooring: Marmoleum Dual (courtesy Forbo Flooring)*

PVC and vinyl

PVC flooring is available as:

a   homogeneous and heterogeneous flexible unbacked polyvinyl chloride, BS EN 649. This can be supplied as a roll or tiles and is bonded to the floor using rubber or acrylic adhesives. It has a high PVC content and can be welded to form a seamless joint

b   jute- or polyester-backed polyvinyl chloride and polyvinyl chloride on polyester felt with polyvinyl chloride backing, BS EN 650. Rolls of backed PVC are attached to the floor using rubber or acrylic adhesives. They have a high PVC content and can be welded to form a seamless joint

c   flexible polyvinyl chloride with polyvinyl chloride foam layer, BS EN 651

d   flexible polyvinyl chloride with cork-based backing, BS EN 652

e   expanded (cushioned) polyvinyl chloride, BS EN 653

f   semi-flexible polyvinyl chloride tiles, BS EN 654. The tiles are bonded to the substrate using a solvent-based bitumen adhesive. They have the highest filler content of all PVC flooring and are consequently the most rigid

g   tiles on a base of agglomerated composition cork with a polyvinyl chloride wear layer, BS EN 655.

BS 8203 provides guidance on how to install all these types of PVC flooring on a wide variety of substrates.

PVC flooring is widely used in public areas where there is a low risk of slipping, and in healthcare, schools and commercial applications. They are also found in domestic use.

PVC flooring is likely to have a poor slip resistance in wet or contaminated conditions.

Surface finishes that incorporate grit improves the slip resistance of PVC flooring (Taylor, 1998). However, in order for some such materials to reach their expected level of slip resistance, a film on the surface of the product must be removed by initial honing, subsequent traffic and cleaning. Its slip resistance may also vary according to direction, being lower in the direction of production (HSE, 2004a). Repeated cleaning and polishing also affect its slip-resistance (HSE, 2004b). For example, a smooth vinyl is likely to loose slip resistance and become smoother; one with a grit through its thickness is likely to show improved slip resistance; and one where the grit is only applied to the surface is likely to lose its slip resistance and become much smoother. Given the variety of possible responses it is important to monitor the flooring's slip resistance and surface roughness over time so as to track changes in its performance.

Figure 10.11 *Vinyl flooring: SureSTEP (courtesy Forbo-Nairn)*

Rubber

Rubber flooring is manufactured from natural or synthetic rubber in sheet and tile, in the following forms:

- rubber with foam backing, BS EN 1816
- homogeneous and heterogeneous smooth rubber, BS EN 1817
- homogeneous and heterogeneous relief rubber, BS EN 12199.

BS 8203 provides guidance on how to install all these types of rubber flooring on a wide variety of substrates.

Figure 10.12 *Profiled rubber flooring on steps, with contrasting stair nosings (courtesy HSE)*

The flooring is fixed to the substrate using a contact adhesive. It provides a quiet and flexible walking surface and is often used in concourses, commercial, hospital and laboratory applications.

Most rubber flooring materials have a relatively low wet slip resistance, although some are available with an acceptable slip resistance. A surface profile of circular studs is common, although this is not thought to increase slip resistance (HSE, 2004b). These studs tend to be worn smooth under heavy traffic conditions (Taylor, 1998). When assessing an existing floor, the surface roughness of rubber flooring should be taken from the surface of these raised circular studs (HSE, 2004b), which will give a useful indicator of the floor's likely performance.

Pendulum testing can be undertaken by an experienced operative to determine the slip resistance of rubber profiled surfaces.

10.6.5   Cork

Cork tile and sheet flooring are made from blocks of compressed granular cork heated with binders. A surface sealer may be applied to prevent the cork surface from becoming impregnated with dirt. BS EN 12104 gives the general requirements for cork flooring.

It is quiet underfoot with good sound absorption. It is susceptible to impact and abrasion damage, attack by most acids and alkalis, and is unsuitable for heavy wear applications.

It is used in domestic applications, places of assembly, libraries, schools and offices, and is usually sealed.

Unsealed cork has good slip resistance, but poor wear characteristics and it is difficult to keep the surface clean. The application of a sealer dramatically lowers its slip resistance in wet conditions.

Figure 10.13   *Cork flooring (courtesy HARO)*

10.7   RIGID FLOORING

10.7.1   *Wood*

Solid wood and pre-engineered wood flooring are increasingly being used in domestic and some office applications. They have also been used extensively for factory flooring because of their hardness and resistance to chemical attack and damage. Strip flooring is common where acoustic and sprung or semi-sprung flooring is required, such as dance halls, sports halls and gyms. Solid wood flooring may be in the form of board strip, block, mosaic, overlay and parquet. Real wood and plastic laminate flooring and panel products are becoming very popular in the domestic market.

Board and strip flooring either uses straight-edged or tongue and grooved boards nailed to, and spanning across, timber battens. Block flooring is laid in an interlocking pattern and bonded to a screed base, often with a bituminous adhesive. Mosaic flooring comprises panels of squares assembled from softwood or hardwood fingers held together by a removable surface membrane or fixed flexible backing. Overlay flooring consists of pre-finished panels or interlocking wood strips of proprietary manufacture for direct application to a fully supporting base. Parquet flooring is made from strips of wood for direct application to a fully supporting base. Laminate flooring comprises a veneer of timber bonded to a particleboard backing. This is cut into tongue and grooved strips and laid as for solid board flooring. BS 8201 is a code of practice for timber flooring and provides guidance on the design and selection, installation and maintenance of timber floor finishes.

Timber flooring tends to be sealed with a polyurethane, melamine or acrylic varnish to prevent the wood becoming permanently soiled. It also reduces differential wear of the surface and protects the edges from moisture penetration. Wax and oil finishes are also used. Further information on wood finishes can be found in TRADA's report 2/2001 (Kaczmar, 2001).

Figure 10.14   *Parquet flooring in a sports hall (courtesy HSE)*

The grain of unsealed timber means that roughness, and hence slip resistance, is likely to vary with direction and species type. Wear may vary from 5 microns with the grain to 50 microns across the grain (HSE, 2004a). Any applied finish is likely to reduce the surface roughness and slip resistance. The SRV in the dry is more than 50, and in the wet is 20–40 (Pye and Harrison, 2003). This does not apply for wax polished floors, which can be very slippery even when dry. Polishes should not be used where it is important that the slip resistance of the timber floor is not altered, such as in gymnasia and sports halls (Kaczmar, 2001). Further investigations by TRADA on the slip performance of more than 30 timber floor finishes on maple and oak showed that none achieved the necessary slip resistance for their use in sports halls in wet conditions (Kaczmar, 2002). Anti-slip nosings should be fitted to timber stair treads (Pye and Harrison, 2003; Taylor, 1998).

10.7.2   *Timber decking*

Timber decks are sometimes considered to have a low slip resistance when wet, but this is most often the result of bad detailing rather than any inherent property of the material. It is essential to prevent a build-up of water on the deck, even a thin layer. An essential aim of the deck construction should be to ensure free drainage of water from the walking surface. Surface water remaining for prolonged periods is likely to cause growth of algae, which are slippery when wet. Frost can also cause timber surfaces to become extremely slippery, as water present in the timber can freeze on the surface.

Most decks are made with planks of solid wood, either of a naturally durable species (and with non-durable sapwood excluded) or suitably preserved softwood. The planks will generally be between 100–200 mm wide, spaced apart to provide drainage. The gap between the planks should be at least 6 mm to avoid either capillary bridging or build-up of dust and debris, with a maximum width of about 12 mm depending on the usage of the deck (football boots or stiletto heels!). To eliminate any twist on planks of 90 mm width or more, two fixings should be used at around the quarter points of the width.

Most timber undergoes residual drying after sawing, which tends to "cup" the surface slightly. If planks are set cup-up they will retain water, as shown in Figure 10.15. In many contracts it may be considered impractical to specify that all planks should be set cup-down, so for planks 100 mm wide or greater it is usual to groove the surface, thereby improving grip for foot traffic and avoiding a build-up of water on the contact surface. Grooving profiles may vary, but they should be at least 6 mm deep to be effective (Figure 10.16). It is prudent to have the decking tested to verify that the grooving is indeed performing as it should. For it to be effective in improving slip resistance, the grooving must be perpendicular to the predominant direction of travel across the decking. If there is no main direction of travel a solution that is effective in all directions should be considered.

Figure 10.15   *A deck of hardwood planks in Paris immediately after heavy rain. The planks set cup-up are retaining water (courtesy Peter Ross)*

Figure 10.16 *Grooved timber decking (courtesy Deborah Lazarus)*

Many decks constructed in accordance with the above recommendations have proved satisfactory in use. Where the deck is large in plan or where the surface slopes (eg at the approach to a footbridge) it is recommended that the slip resistance provisions should be upgraded in one of the following ways.

1 An overlay of chicken wire (Figure 10.17). This is a cheap solution, but requires fasteners at frequent intervals. It is effective in all directions of travel. The mesh is vulnerable to damage and then appears unsightly. Depending on the surrounding area, the mesh may become clogged with mud and debris in very wet conditions, reducing its effectiveness significantly. It is therefore not maintenance-free.

Figure 10.17 *Timber decking with chicken wire overlay on footbridge (courtesy Peter Ross)*

2 Non-slip inserts. The type shown in Figure 10.18 is a sand-epoxy mix, applied like a sealant into a dovetail groove, which holds it in place when set.

3 Small battens fastened to the surface.

Figure 10.18 *Sand-epoxy strips set into a dovetail groove (courtesy Peter Ross)*

The design and detailing of timber decks is dealt with more fully in the TRADA *Timber decking manual* (TRADA Technology, 1999).

Ceramic

Ceramics are made from refined clays. A wide range of ceramic flooring materials is available, ranging from the rough (clay pavers) to the very smooth (ceramic glazed tile). They are classified in BS 6431-1 and fall into two main classes according to their method of manufacture. Extruded tiles are shaped in the plastic state in an extruder and the resulting column of clay cut into tiles of predetermined thicknesses. Dust-pressed tiles are formed of powder or small grains, which is shaped in moulds under high pressure before firing. The latter are generally made to finer dimensional tolerances than extruded tiles. Further details of different ceramic products are given in The Tiling Association's guide, *Slip resistance of hard flooring* (TTA, 2002).

The extent of firing determines their porosity and hence their strength and ability to absorb water. Fully vitrified tiles have a water absorption of less than 0.5 per cent by weight. They are available in different thicknesses and sizes and may be glazed or matt and can be given a surface profile (HSE, 2004b). They may be adhesively bonded to a substrate or laid unbonded, for example bedded in sand.

Ceramic tiles are widely used in shops, toilets, food preparation areas and reception lobbies because of their longevity, cleanability and appearance.

Unglazed floor tiles are not usually slippery when clean and dry, but, as with all flooring materials, if water, dust, oil, grease or wax is present on the surface, potentially slippery conditions will be created. Flood-grouting of ceramic floors may have a detrimental effect on the slip resistance as particles of grout reduce the ex-factory roughness. Ceramic tile surfaces exhibit roughness values that range from very low (less than 5 microns) to relatively high (around 20 microns), dependent on type and glazing finish. As is the case for all floor finishes, their surface finish, slip resistance and surface roughness need to be appropriate for the use. Fully vitrified glazed tiles may be very smooth and so only suitable for dry, well-maintained areas, whereas tiles containing an abrasive grit will have an increased slip resistance (Taylor, 1998) and be suitable for use in areas subject to more viscous contamination (TTA, 2002). The slip resistance of ceramic tiles alters as they wear. Highly trafficked areas may become worn appreciably faster than areas that are more lightly used. Where slippery conditions in service may present a significant hazard, especially on steps and where floors are laid to steep falls, tiles or inserts with slip-resisting finishes should be used.

Slips in kitchen on ceramic tiles

A kitchen worker fractured her skull in a slip incident in a store restaurant. Four other slip accidents had occurred in the preceding 12 months.

The employer was ordered to pay £36 000.

Inspection showed the tiles felt slippery, although they were in good condition and appeared clean and dry. When contaminated with small quantities of water or grease it became extremely slippery. Some areas were sloped, making matters worse. The "safety" mats that had been installed in some areas (which were themselves unacceptably slippery) had been removed by cleaners at the time of the accident and the floor was contaminated with water and oily residue. The floor surface was eventually replaced with one that was suitable for use in an area where the total elimination of floor contaminants would never be possible. The new floor was specified to provide enough grip, even in wet or contaminated conditions.

Source: HSE, <www.hse.gov.uk/slips/experience/skullfracture.htm>, accessed 1 Dec 2005

Porcelain, fully vitrified, vitrified and semi-vitrified

These tiles may be glazed or unglazed. The glaze on a ceramic tile is made from a glass that melts during firing and forms a smooth hard layer on the surface of the tile. Glazed tiles are widely used in domestic applications in bathrooms and kitchens on walls and floors.

A surface roughness of less than 5 micron is typical for glazed tiles (Lemon, Thorpe and Griffiths, 1999a). Glazed tiles should not be used in areas likely to become wet unless they designed to be slip-resistant. When it is known that in-service slippery conditions may arise and present a significant hazard, especially on steps and where floors are laid to steep falls, tiles or inserts with slip-resisting finishes should be used (BS 5385-3). Floor surfaces may become slippery in time through the polishing action of traffic.

Figure 10.19 *Fully vitrified ceramic granite tactile tiles containing corundum anti-slip additive, with layout to BS 7997 (courtesy Shackerley (Holdings) Group/The Tile Association)*

Quarry tile and terracotta

Quarry tiles are cut from a single clay extrusion. They may be pressed subsequently or made individually. Typically, they are terracotta, cream or black in colour and are not glazed.

Terracotta tiles tend to be made from natural clays and have a water absorption of greater than 10 per cent.

Quarry tiles are widely used in many industries, both externally and internally, but especially in the catering and leisure sectors in kitchens and food preparation areas.

Figure 10.20 *Quarry tile flooring in catering area (courtesy HSE)*

The finished tile may be sealed, but this is not recommended because the slip resistance and cleaning characteristics can be changed dramatically. Aggregates can be added to quarry tiles to enhance the slip-resistance characteristics of the wearing surface. A typical aggregate is carborundum grit, which may be dispersed throughout the tile matrix or applied to the wearing surface only. Those tiles with grit applied to the surface only will wear with use and the initial characteristics will change accordingly.

Tiles with added aggregate may exhibit slip-resistance levels two to three times that of standard quarry tiles and have a surface roughness of 25–40 microns. They should be quite capable of combating the majority of common workplace contaminants expected in toilets, kitchens, food counters and bars.

10.7.6   Clay brick paving and clay pavers

These are brick-sized clay paving blocks or clay pavers bedded in concrete or sand to form a uniform pavement.

Clay pavers are used in industrial and commercial applications, usually in external applications.

Their surface texture may vary from very rough to relatively smooth. Some brick paving, such as those of a hard-burnt stock brick type, have a rough texture that provides a good slip-resistant finish. Certain facing bricks may provide similarly good slip resistance properties when laid on edge, and some will be suitable when laid on the bed face. Some hard clay brick paving produced by the wire-cut process have a resultant rough micro-texture that provides excellent frictional resistance and are suitable for applications where the pavement may be subjected to vehicular traffic. Smoother surfaces may be present on engineering bricks or similar dense pavers, although some purpose-made clay pavers of this type incorporate an artificially patterned surface with a panelled or chequered effect. These smoother, dense pavers may be well suited to internal heavily trafficked pedestrian areas, such as shopping arcades (Harding and Smith, 1995).

Clay pavers and brick paving tend to have good slip resistance in wet and dry conditions because of their initially high surface roughness values (Taylor, 1998). According to BS EN 1344, clay pavers are considered to have satisfactory slip/skid resistance during the working life of the product provided they undergo normal maintenance and have not been subject to grinding and/or polishing to produce a very smooth surface.

BS EN 1344 gives a series of classifications for the unpolished slip resistance of clay pavers used in flexible paving, measured using the pendulum test and with a standard rubber slider. The equipment is equivalent to that specified in BS 7976-1. However, it should be noted that the CEN rubber slider used in the standard is not usually recommended as an alternative to the Four-S or TRL rubber sliders (UKSRG, 2005):

U0 no determination

U1 SRV ≥ 35

U2 SRV ≥ 45

U3 SRV ≥ 55

Although these classifications exist, they are not referred to in any current guidance on acceptable levels of slip resistance, except that a U3 clay paver may be advisable where it is to be used on an inclined surface such as a ramp.

According to the HSE guidance, clay pavers with classifications U2 or U3 have low potential for slip if measured using Four-S rubber (see Table 3.3). The majority of pavers complying with U1 would have a low potential for slip, unless the SRV was 35, which would be classified as having moderate potential for slip.

Figure 10.21 *Clay brick paving near the Trafford Centre, Manchester (courtesy Tony McCormack, <www.pavingexpert.com>)*

Various types and forms of stone are used as flooring both internally and externally. Widely used stones include sandstone, marble, quartzite, travertine, granite, slate and limestones. Their finish may be very smooth, as is often the case for polished granites or marble in office foyers, or very rough, in external paving and setts, for example. BS 5385-5 covers the design and installation of stone floors and BS EN 1341 covers external stone paving. The CEN rubber slider used in the standard is not usually recommended as an alternative to the Four-S or TRL rubber sliders (UKSRG, 2005).

As stone is a natural material it is highly variable and its selection will depend on its mechanical properties, appearance, bedding direction, jointing and surface finish. Different stones will wear at different rates.

Stone tends to be used in high-value areas such as entrance halls to offices, banking halls, lobbies and retail concourses. All these areas are foreseeably wet.

Figure 10.22   *External stone paving (courtesy HSL)*

Figure 10.23   *Wath Blue limestone at Fetter Lane, London EC4 (courtesy Cadeby Stone Limited)*

According to BS EN 1341, coarse-textured and riven slabs are assumed to give satisfactory slip resistance for external stone paving. However, many riven surfaces are not as slip-resistant as their appearance suggests. The undulating surface typical of riven materials has little additional slip resistance and is often relatively smooth and becomes slippery when wet. An experienced pendulum operator can obtain valid data from such surfaces. They may also be tested using ramp-based methods.

Applied polishes and varnishes are not recommended, as they reduce slip resistance in the wet.

BS EN 1341 requires that the producer declare the minimum unpolished slip resistance value (USRV) expected for individual test specimens of fine-textured slabs when tested with a pendulum friction tester, as described in the standard. The standard does not give a minimum USRV requirement.

BS EN 1342 for setts of natural stone requires the same, but also notes that the performance of setts when laid may have a different slip resistance value to that determined on individual setts or test specimens.

BRE undertook slip-resistance testing on 34 types of UK stone used in flooring (limestones, sandstones and slates) according to BS EN 1341 on wet specimens (Yates and Richardson, 2000). They drew the following conclusions for wet conditions:

- stones with a honed or sawn finish usually have a SRV > 70
- stones with a more polished finish may have a SRV < 40
- the smoother the stone, the more likely it is to be slippery
- stones with a polished finish have a surface roughness of less than 2 ?m
- rougher stones may become slippery if the surfaces of individual grains becomes polished.

It should be noted that stones can be honed to have a very wide range of SRV and testing has indicated that they do not necessarily achieve a wet SRV of more than 70. Consequently, testing should be undertaken on the actual stone and finish to be used.

10.7.8 Concrete paving

Concrete paving is produced using either wet cast, vibrated or semi-dry methods. Standard concrete paving flags are typically hydraulically pressed.

It is widely used for external pedestrian areas and pavements.

BS EN 1338 and BS EN 1339 state that concrete paving blocks and flags have satisfactory slip/skid resistance provided their whole upper surface has not been ground and/or polished to produce a very smooth surface. If the surface of a block contains ridges, grooves or other surface features that prevent testing by the pendulum friction equipment, the product is deemed to satisfy the requirements of this standard without testing. This is unsatisfactory, as surface features do not necessarily add slip resistance to any surface and it is prudent to test any surfaces with these features. An experienced pendulum operator can obtain valid data from these surfaces. They may also be tested using ramp-based methods.

BS 6717 states that precast, unreinforced paving blocks should have their SRV measured using a pendulum tester that complies with BS 812-114. BS 7263 states that precast concrete flags, kerbs, channels, edgings and quadrants should be tested in the same way and should comply with the requirements below. (BS 812-114 has been superseded by BS EN 1097-8, which is also a test method for assessing slip resistance of aggregate.) According to BS 6717 and BS 7263, the paving block should be tested in the unpolished and polished condition, and the lowest value taken as the SRV of that paving. The samples are soaked in water before testing. The standard classifies paving units as S1, S2, S3 and S4:

- S1 = no performance determined
- S2 ≥ 35
- S3 ≥ 45
- S4 = manufacturer's declared value.

The standards state that blocks complying with S2 are suitable for use in pedestrian areas and S3 for vehicular access areas.

Other standards also address the slip resistance of concrete paving units:

- BS 7923
- DD ENV 12633
- BS EN 1338
- BS EN 1339.

None of the standards above appears to have been superseded and it is not clear which is the most appropriate to use.

Figure 10.24 *Concrete paving units (courtesy Deborah Lazarus)*

10.7.9 Tactile paving

Following extensive research with user groups having a variety of disabilities, a range of tactile paving was developed to provide essential warning and other information to vision impaired pedestrians. In the UK there are seven types of tactile surfaces (BS 7997).

1 Blister surface, to warn of a dropped kerb or that the road has been raised to the height of the kerb.

2 Platform-edge warning surface (off-street trains and trams), to warn people that they are approaching the edge of a railway platform.

3 Platform-edge warning surface (on-street trams and rail transport), to warn people they are approaching the platform edge at a tram stop.

4 Corduroy hazard warning surface, giving the message "hazard, proceed with caution", commonly used at the top and bottom of stairs and ramps.

5 Tactile surface at shared-use routes to inform vision-impaired pedestrians that they are entering this type of area, that they are walking on the pedestrian and not the cyclists' side, and to help them keep to this side of the facility. This surface is used in conjunction with a raised trapezoidal profiled line marking.

6 Guidance paths, to help people negotiate large open spaces.

7 Information surface, used to highlight information points such as ticket offices.

Tactile surfaces can be detected by vision-impaired people because they have raised profiles, which can be felt underfoot or with a cane. The information surface is the exception to this and is detected by being softer than the surrounding paving material. In order for these surfaces to be read correctly by a visually impaired pedestrian they must be installed in a consistent manner. The consistent approach applies not only to the types used, but also their location and orientation with respect to the hazard. Research by HSL has shown that tactile paving is often not installed correctly (Loo-Morrey, 2005b).

Figure 10.25 *Incorrectly installed corduroy warning strip– too narrow and not in a contrasting colour (courtesy HSL)*

The majority of vision-impaired people can detect light from dark, so tactile surfaces should contrast in colour and tone with the background. This contrast also helps highlight their presence to other pedestrians. The contrast can be affected by street lighting conditions (UCL, 2005).

Guidance on the use of tactile paving in the outdoor environment is contained in a document available from the Department for Transport (DETR, 1999). Approved Document M (ADM) requires that tactile paving should be used on access routes to provide warning and guidance to vision-impaired people.

A tactile surface will not, in itself, affect the SRV of the flooring material – the SRV will depend on the material and surface finish used. Most manufacturers create surfaces from established footway materials that often have a reasonable SRV, including concrete and stone. However, in recent years other means of creating the blister surface have included forming the blister shape from brass or other metals, and punching these into the pedestrian surface. The SRV of materials such as metals and plastics should be checked to ensure they do not decrease slip resistance.

Research by TRL Ltd for the Rail Standards and Safety Board (RSSB) (Sentinella *et al*, 2005) has considered whether tactile surfaces in the railway environment, particularly at the platform edge and corduroy warning surfaces, create additional risks to other pedestrians. The research showed no reliable evidence to suggest that tactile surfaces pose a risk to passengers. However, they were found to prevent accidents to vision-impaired pedestrians at the platform edge.

10.7.10 Glass

Glass floors comprise panes of glass mounted in a steel frame. The glass is typically toughened and laminated and the thickness will depend on the size of the pane. A frit or sandblasting is frequently applied to the upper or lower surface to reduce transparency from below, to provide a decorative pattern or to impart some surface roughness (if applied to the upper surface).

Floors using glass are becoming increasingly popular in top-of-the-market applications, in retail, domestic and office premises. Glass may also be used in pedestrian surfaces for covers to lighting or other inserts.

Because of its smoothness (less than 1 micron) glass can be extremely slippery when wet and should not be used untreated for flooring or stair treads. However, techniques that enhance its surface roughness such as etching and grit-blasting can give it reasonable slip resistance in wet conditions. Some of these techniques do, however, cause the glass to lose its transparency (HSE, 2004b). Designers should seriously consider the slipping risks of untreated glass before opting for walked-on glass flooring (HSE, 2004b).

Sudden transition from opaque surfaces to transparent surfaces can also be a problem to pedestrians with vertigo, especially if there is a significant difference in perceived height. Where glass is used as an external roofing material laid to fall for drainage purposes and is expected to be walked on either as part of a traffic route or for access for maintenance and repair, even higher slip resistance is required to offset the extra demands of the incline.

Figure 10.26   *Glass floor with fritted surface (courtesy HSL)*

10.7.11 Metal

The most common form of metal flooring is steel or aluminium profiled sheets or grids (HSE, 2004B). Typically it is mechanically fixed to the substrate.

Figure 10.27   *Steel "Durbar" patterned metal flooring (courtesy Deborah Lazarus)*

Metal flooring is common in industrial situations, particularly in food processing and manufacturing. Other uses include loading bays, step-ladders and ambulance floors. It is also becoming fashionable as a floor finish in retail outlets.

While the presence of the profile may give some mechanical interlocking with a cleated shoe heel or sole, this is not necessarily the case. Where interlocking does not occur, eg with smooth shoe heel and sole surfaces, the surface roughness of the top of the profile determines the overall slipperiness of the surface. Metal flooring is thus often much more slippery than anticipated. Users typically perceive a profiled floor surface as less slippery than a flat surface. Lulled into a false sense of security, they often take less care than they would on a surface they perceive as slippery, potentially leading to higher rates of slipping. HSL research indicates that, in contaminated conditions, both smooth and profiled aluminium sheet have high slip potential. Smooth steel flooring also has a high slip potential, but the addition of a profile can significantly reduce slip potential. Mild steel generally has better slip resistance than aluminium because it is harder and tends not to wear smooth. Also the surface roughness increases as the steel rusts.

Profiled surfaces will wear with use and any slip-resistance properties gained by galvanising or other anti-slip coatings will change.

10.8 ACCESSORIES

There are numerous accessories associated with flooring, including movement joints, drain covers, gulleys and gratings for use in highly contaminated areas. The most important are discussed below.

10.8.1 Gratings

The slip resistance of 15 floor grating panels was studied by the HSL ramp test (Lemon *et al*, nd). They were steel, ceramic or GRP. In some samples the aluminium paint rapidly wore off the steel grids and the SRV decreased. This would suggest that the in-service performance of a worn grating, where the abrasive layer had been lost, would have a much lower slip resistance. In one of the GRP gratings some carborundum grit detached from the surface during testing. This significantly reduced the SRV, and similar behaviour could be expected in service, calling the durability of this grating into question. However, testing showed that the SRV in wet conditions was more than 40, so they would provide adequate slip resistance in the test conditions.

Two factors suggest that although the SRV is adequate in the tests, it may well be lower in practice:

- the contaminant was clean water; in service it could be a more viscous contaminant
- the surface treatment of some gratings wore off quickly. This was found to be the case through experience of similar flooring materials in a food production plant under viscous contamination conditions.

Figure 10.28 *Metal grating*

Escalators are common in shopping centres, stations and other public areas. Their treads and the landing immediately adjacent to them are usually steel. The landings may have a profiled surface (Figure 10.29). As discussed in Section 10.4, the tops of such profiles frequently become worn and slippery, so although the profile gives an appearance of safety, this is not necessarily the case. Wear rates are likely to be high at the head and foot of an escalator and the landing surface, as pedestrians tend to step in the same place. This can lead to accelerated wear (Figure 10.30) and higher slip potential, particularly in the presence of unexpected contamination.

Figure 10.29   *Profiled metal surface at head of escalator (courtesy Deborah Lazarus)*

Figure 10.30   *Wear immediately adjacent to escalator steps (courtesy Deborah Lazarus)*

Others

Expansion joints, floor jointers, socket covers, air intake covers and other accessories for all types of floors should exhibit similar slip resistance properties to that of the main floor. Many slip accidents are caused by a sudden change in slip resistance characteristics from one surface to another that is not immediately obvious to the pedestrian.

Figure 10.31 *Polished brass air intake duct cover to carpeted raised floor (courtesy HSL)*

The polished brass air intake duct cover shown in Figure 10.31 has low slip resistance compared with the surrounding carpet; if contaminated it will have a significant slip potential.

Figure 10.32 shows a metal gulley cover installed in front of an entrance. The adjacent material is stone or paving and would have a much higher slip resistance than the gulley in wet conditions.

Figure 10.32 *Metal gulley cover at threshold (courtesy HSL)*

Glass pavers are widely used in external paving. Glass light covers are also increasingly incorporated in pedestrian surfaces. All such glass needs to have good slip resistance.

The glass lights in Luxcrete paving are either in the smooth as-pressed condition, or have been sand-blasted and a clear liquid polymer coating applied to protect the glass before the concrete is cast around them. In as-pressed condition, the glass is initially smooth compared with the concrete, but wear from foot traffic quickly gives it an opaque finish, which is rougher. For the coated lights, the polymer coating soon wears away in service, leaving the sand-blasted glass surface exposed. The Greater London

Council chose the maximum size of glass (100 mm × 100 mm) in the 1930s specifically to minimise risk of slipping. In addition, grit is added to the concrete surface to enhance slip resistance.

Testing undertaken on behalf of Luxcrete showed that while the SRV of the glass is always lower than the concrete, the values are sufficiently high to provide adequate slip resistance in wet conditions. Pendulum testing indicted that the mean SRV in wet conditions was 45 for the glass (sand-blasted and lacquered, representing the newly installed treated blocks) and 72 for the concrete. In dry conditions these values increased to 76 and 87 for the glass and concrete respectively.* Once the lacquer wears away, the slip resistance of the glass probably increases slightly, because the coating is likely to reduce the surface roughness. Similarly, the as-pressed glass will have lower slip resistance when it is installed initially until it becomes worn.

Figure 10.33 *New (courtesy Luxcrete Ltd) and in-service Luxcrete pavement lights (courtesy K Brassington)*

10.9   SURFACE ALTERATION TECHNIQUES

Several techniques exist that purport to increase the slip resistance of flooring. Although there is overlap in methods, these should not be confused with those techniques used to prepare an existing surface before a new floor finish is laid – scabbling, shot-blasting etc. Techniques to improve slip resistance vary with the material concerned and are usually limited to hard surfaces such as concrete, ceramics and metal flooring, and include the use of acid-etching, grit-blasting, dry-shake additives, the application of slip-resistant tapes or sheets and "slip-resistant" floor polishes. Their efficacy has been investigated and further details are given below (Lemon, Thorpe and Griffiths, 1999b).

10.9.1   *Acid etching*

Acid-etching is commonly used as a way to increase the surface roughness of hard floor finishes such as ceramic, natural and man-made tiles. It is often employed as a result of an inappropriate flooring choice given the level and type of contamination expected. Highly polished, low-roughness, fully vitrified ceramic tiles in bathrooms, toilets and kitchens are examples.

---

*   The above test results relate solely to Luxcrete products and none other.

Four acid-etch treatments were investigated (Lemon, Thorpe and Griffiths, 1999b). Because of commercial sensitivity, the exact nature of the treatments is not stated, beyond the fact that hydrofluoric acid was often the main active ingredient of the etch solution. The manufacturers applied the treatments on to panels of fully vitrified polished floor tiles and glazed ceramic floor tiles. DIN ramp testing, pendulum testing and their roughness were compared before and after testing. The key findings were:

- changes in surface roughness were less than 1 micron in size, and below the detection limit of the surface profile measurer. The increase in surface micro-roughness was evident in the surface reflectivity and scratch tests on the surface

- all four treatments gave a significant increase in slip resistance for the wet shod condition, but only three gave a similar increase for the barefoot condition. The increase in performance was more pronounced for the glazed floor tiles compared with the vitrified ones, because they had much lower slip resistance to start with.

Surface alteration by acid-etching is effective in increasing surface roughness and so improving slip resistance. Yet no matter how advantageous this is in combating the potential for a slip accident this must be balanced with the following negative attributes:

- virtually all acid-etching techniques involve the use of hydrofluoric acid (HF) or ammonium bifluoride and its derivatives – a particularly dangerous acid to the health of the applicator. HF is a poison that accumulates in the bones and causes weakening and degeneration of the skeletal structure. HF vapours can dissolve in the moisture of the eye and cause irritation. HF burns are extremely painful, eating through tissue and attacking the bone

- the cleaning solutions sold as part of the etch treatment also contain HF. Cleaners require proper training and equipment in order to use them

- etching solutions may attack grout in tile joints and penetrate beneath the tiles through poor grouting to attack adhesives and sub-floor integrity

- warranties from flooring suppliers are likely to be compromised or even negated

- the expected durability of the treated surface may be significantly shortened

- the cleaning regime will require rethinking to ensure it is capable of cleaning a floor with changed roughness characteristics.

Acid-etching of museum floor

A prestigious new museum, due to open shortly, was found to have unacceptably slippery flooring. Designers had focused on the aesthetics of the floor, but not its slip resistance. Testing showed that the floor was extremely slippery under water contamination. This was of particular concern because some areas of the floor were in bar and café areas.

To identify a suitable surface treatment, laboratory testing was undertaken to determine the effectiveness in increasing the slip resistance value and the surface micro-roughness on the installed floor for a number of acid etches. A suitable proprietary acid-etch treatment was selected and successfully applied on all foreseeably wet areas.

Further on-site testing carried out after treatment showed a large reduction in the slipperiness of the floor surface in all areas with high slip potential. The treatment incurred significant cost, changed the appearance of the floor and probably reduced the lifetime of the floor considerably. Correct initial specification of the flooring would have been far more effective.

Subsequent visits showed that the etched areas were harder to clean with the existing regimes, leading to a build-up of unsightly oil.

Source: HSE, <www.hse.gov.uk/slips/experience/museum.htm>, accessed 1 Dec 2005

It is far better to get the choice of tile right at the design stage rather than have to resort to the use of etching and/or other surface-altering techniques.

10.9.2   Tapes and sheets

Slip-resistant tapes and sheets are available in a range of grades for application to existing surfaces. The size and profile of the aggregates used may vary, and will influence their slip-resistance. If the aggregates are rounded as opposed to sharp, the slip resistance is slightly lower. Partial detachment may occur under prolonged use, and this will reduce their performance and constitute a trip hazard. The typical SRV of the tapes tested under wet conditions was 90–100 (Lemon, Thorpe and Griffiths, 1999b). It is important that the surface to which any anti-slip tape is fixed is clean, dry and free from dust. Floors need to be degreased and the surface to which the tape is to be applied should be relatively smooth. Porous surfaces, such as untreated wood and concrete, may have to be sealed before application.

Tapes should be spaced so as to minimise the amount of potentially dangerous flooring. They should be sufficiently close together to make it difficult to step between them. Where applied to the nosing of a stair they should be applied as close to the leading edge of the nosing as to minimise or eliminate the existing finish. Figure 10.34 shows retro-fitted anti-slip tape fitted to old unpolished terrazzo stair treads. The tape is too far away from the leading edge of each step to be fully effective and the whole detail made much worse by the existing bull-nose nosing. In this particular case it would have been more appropriate to affix the tape to the curve of the nosing itself.

Figure 10.34   *Incorrectly located anti-slip strip applied to stair goings (courtesy HSL)*

10.9.3   Floor polishes

Tests were carried out on standard and "anti slip" floor polishes applied to glazed and fully vitrified ceramic tiles, according to manufacturers' recommendations (Lemon, Thorpe and Griffiths, 1999b). The application of standard and anti-slip polishes to vitrified or glazed ceramic floor surfaces did not appear to affect the slip resistance to any appreciable degree. Rapid delamination of polish from the test surface suggested that the life of such polishes may be very short.

HSE also tested the influence of floor polishes on heavy-duty vinyl flooring. DIN and pendulum testing showed that the application of polish reduced the slip resistance of the surfaces. In some cases, the surface roughness of the vinyl also decreased by more than 75 per cent (Lemon, Thorpe and Griffiths, 1999b).

Rotary buffers may be used in the cleaning of floors and the progressive effect they have on smooth vinyl floor was investigated. A progressive increase in slip potential and decrease in surface roughness were observed. When tested wet, the slip potential increased from just in the low range to moderate within six application cycles. If further applications were made, it is likely the slip potential would have become high (Lemon, Thorpe and Griffiths, 1999b).

DATA ON WALKING SURFACE MATERIALS SLIP RESISTANCE

HSE caution on tabulated values

Table 10.1 (on pages 158–163) contains expected roughness and slip resistance values of different flooring types obtained largely from laboratory-based tests on factory-gate samples. While the figures shown are those as found, they are based largely on manufacturers' samples and do not indicate the effects of laying, conditioning, cleaning and wear associated with use. All generic materials will have a range of results because of material variation. For example, terrazzo can vary from 5 microns to 65 microns depending upon aggregate type and finishing; for timber there may be variations in across-the-grain and with-the-grain values as well as between species.

All flooring materials change with use – some become smoother, others rougher. Specifiers of flooring materials should verify manufacturers' claims for new materials and how they will expect them to change over the life expectancy of the floor. The testing of similar types of *in-situ* floors subjected to cleaning and wear can increase confidence in product claims.

Using a roughness meter in conjunction with the Slips Assessment Tool will produce a guide to a floor's performance. Accurate measurements of the floor's SRVs should be taken using the protocol agreed by the UK Slip Resistance Group. Very rough floors, eg resin, quarry and vinyl safety floors with added grit particles, may produce anomalies leading to lower roughness values due to the diamond stylus of the meter being diverted around the particle.

The performance of products with Rz values outside the typical roughness ranges may differ from that expected and their properties should be determined using pendulum testing. Where a floor surface is stated to have a high slip potential, this refers to contaminated surface.

Table 10.1   *Typical data on walking surface materials slip resistance based on HSL experience*

Flooring material	Finish	Slip potential in wet conditions			Typical roughness (value in µm)	Expected SRV, dry/wet	Comment
		Low	Moderate	High			
Concrete							
C1 Concrete	Floated and/or brushed or similar	–			50	75/65	Rough floor surface will combat majority of expected contaminants. Contaminants likely to be absorbed into floor surface increasing slip potential.
C2 Concrete	Power float – several years old	–			20	65/40	Safe floor surface when dry and clean but may be compromised by more viscous contaminants such as heavy oils or grease. Contaminants likely to be absorbed into floor surface, increasing slip potential. Floor several years old showing signs of surface dusting but no significant increase in slip potential.
C3 Concrete	Power float – new		~		15	65/35	Finish can be very variable across whole surface depending upon the amount of power-floating. Can be smooth surface with moderate to high slip potential, especially for dry contaminations.
C4 Concrete	Power float with added dry-shake topping			≈	5	75/10	Highly durable easy-to-clean floor, but also very smooth surface. Safe floor surface when dry and clean, but high slip potential when exposed to any common wet or dry contaminant. Floor must be kept clean and dry at all times, with extra vigilance during inclement weather or spillage.
Natural and man-made stones							
N1 Granite	Polished			≈	3	75/10	Highly durable, aesthetically pleasing, easy-to-clean floor with very smooth surface and high slip potential. Safe floor surface when dry and clean, but will become very slippery when exposed to any common contaminant. Floor must be kept clean and dry at all times, with extra vigilance during inclement weather.
N2 Granite	Flamed	–			45	75/45	Highly durable, aesthetically pleasing floor with low slip potential. Far more demanding cleaning regime. Significant roughness and SRV creates much safer floor when exposed to any common contaminant. Can be used safely on inclined surfaces both internal and external.
N3 Granite	Bush hammered	–			65	75/48	As N2.
N4 Limestone	Honed			≈	8	75/10	As N1. High slip potential.
N5 Limestone	Polished			≈	10	75/10	As N1. High slip potential.
N6 Limestone	Natural	–			25	85/45	Durable, aesthetically pleasing floor, but with more demanding cleaning regime than polished surface. Higher roughness and SRV creates safer floor when exposed to many common contaminants.
N7 Sandstone	Natural cleft	–			40	85/65	As N5.
N8 Marble	Polished			≈	1	75/10	As N1. High slip potential.

Flooring material	Finish	Slip potential in wet conditions			Typical roughness (value in µm)	Expected SRV, dry/wet	Comment
		Low	Moderate	High			
N9 Terrazzo	Ground and polished finish			≈	3	63/9	As N1. High slip potential.
N10 Terrazzo	Safety finish	–			65	85/65	Highly durable, aesthetically pleasing floor, but with far more demanding cleaning regime than standard terrazzo. Significant roughness and SRV creates much safer floor when exposed to any common contaminant.

Resilient

Flooring material	Finish	Slip potential in wet conditions			Typical roughness (value in µm)	Expected SRV, dry/wet	Comment
		Low	Moderate	High			
Res1 Vinyl	Standard matt finish			≈	14	66/15	Durable and easy-to-clean floor, but with smooth surface. Safe floor surface when dry and clean, but may become slippery when exposed to common contaminants. Floor must be kept clean and dry at all times, with vigilance during inclement weather. High slip potential.
Res2 Vinyl	Standard sheen finish			≈	7	66/10	Durable and easy to clean floor, but with very smooth surface. Safe floor surface when dry and clean, but will become slippery when exposed to common contaminants. Floor must be kept clean and dry at all times, with vigilance during inclement weather or spillages otherwise high slip potential.
Res3 Vinyl	Safety	–			43	56/34	Durable but with far more demanding cleaning regime than standard vinyl. Significant roughness and SRV creates much safer floor when exposed to any common contaminant. SRV will improve with traffic as aggregate exposed. Poor cleaning will increase slip potential. Better-quality floors have grit particles throughout the section.
Res4 Vinyl	Studded		~	≈	5	Test to verify usefulness of profile	Durable and easy-to-clean floor but smooth surface. Matt finishes more slip-resistant than sheen finish. Safe floor surface when dry and clean, but may become slippery when exposed to most common contaminants. Additions of stud patterns may not contribute to better slip resistance. Therefore it has a high slip potential. The surface profile may also make it more difficult to dry.
Res5 linoleum	Standard			≈	5–10	75/10	As R1. High slip potential.
Res6 rubber	Natural			≈		70/20	As R1. High slip potential.
Res7 rubber	Synthetic			≈		70/20	As R1. High slip potential.
Res8 rubber	Studded			≈		Test to verify usefulness of profile	Durable and easy to clean floor but smooth surface. Matt finishes more slip resistant than sheen finish. Safe floor surface when dry and clean, but may become slippery when exposed to common contaminants. Additions of stud patterns may not contribute to better slip resistance. Therefore it has a high slip potential. The surface profile may also make it more difficult to dry.
Res9 cork	Sealed			≈	5	75/15	Aesthetically pleasing and warm floor with better durability and slip potential characteristics. Addition of sealant or polishes will dramatically increase slip potential. High slip potential.

Flooring material	Finish	Slip potential in wet conditions			Typical roughness (value in μm)	Expected SRV, dry/wet	Comment
		Low	Moderate	High			
Timber and laminates							
T1 Timber	Engineered veneer on ply base – unsealed surface	−			15	75/35	Durable, aesthetically pleasing, easy-to-clean floor but with smoothness of surface dependent upon species type, sanding and sealant. Natural, unsealed timbers have directionality characteristics, with cross-grain usually more slip-resistant than with-grain. Safe floor surface when dry and clean, but will become slippery when exposed to most common contaminants.
T2 Timber	Engineered oak veneer on ply base – sealed surface			≈	6–7	64/23	Durable, aesthetically pleasing, easy-to-clean floor, but sealant will reduce surface roughness to very low figures, thereby increasing slip potential. High slip potential.
T3 Timber	Strip or board. Sanded – sealed		~		10	75/35	As T2. High slip potential.
T3 Timber	Parquet – sealed			≈	3	56/19	As T1. High slip potential.
T4 Timber	Laminate			≈	8	60/20	Durable, aesthetically pleasing, easy-to-clean floor. Laminates have some grain directionality dependent upon mould properties. Floor must be kept clean and dry at all times, with extra vigilance during inclement weather. Unsuitable for foreseeably wet areas. High slip potential.
T8 Timber	Deck – grooved profile		~		9	Test to verify usefulness of profile	Durable flooring material, but subject to mould and lichen growth when used externally. Timbers swell with water and SRV may change dramatically from dry to wet state. Grain profile important in traffic flow. Addition of grooves or slip-resistance strips may increase SRV but require verification. High slip potential.
Quarry and ceramic							
C1 Quarry	Standard	−			13	61/54	Highly durable floor material but with range of roughness and SRVs only likely to combat clean water.
C2 Quarry	Safety	−			25	85/45	Highly durable floor finish with added grit significantly lowering slip potential when surface contaminated with oil and grease. Poor or infrequent cleaning will increase slip potential. Better-quality floors have grit particles throughout the section.
C3 Ceramic	Matt finish	−			19	60/40	Highly durable floor material with increased range of roughness and SRVs likely to produce lower slip potential values.
C4 Ceramic	Pressed (without surface modification to enhance roughness)			≈	5–10	85/10	Highly durable and aesthetically pleasing, but many with low roughness and SRV in wet conditions, so have high slip potential. Floor surface needs to be kept clean and dry at all times, with extra vigilance during inclement weather or spillages. High slip potential.

Flooring material	Finish	Slip potential in wet conditions			Typical roughness (value in μm)	Expected SRV, dry/wet	Comment
		Low	Moderate	High			
C5 Ceramic	Profiled			≈	5–20	Test to verify usefulness of profiled surface	Highly durable and aesthetically pleasing but there are many profile patterns associated with low roughness and SRV in wet conditions and high slip potential. Floors with high slip potential need to be kept clean and dry at all times, with extra vigilance during inclement weather. Surface should be tested to confirm SRV, as profile not necessarily beneficial.
C6 Ceramic	Vitrified. Full polish			≈	1	70/5+	Highly durable, aesthetically pleasing, easy-to-clean floor but very smooth surface. Safer floor surface when dry and clean, but very slippery when exposed to any common contaminant. Floor must be kept clean and dry at all times, with extra vigilance during inclement weather. High slip potential.
C7 Ceramic	Surface acid-etched	—			Depends on treatment and application	70/45	Surface increase in roughness dependent upon amount of acid used. 95 per cent of etch methods contain hydrofluoric acid or its precursors. BEWARE – see Section 10.9.1
Metal and GRP							
M1 Aluminium sheet	No profile			≈	3	60/13	Highly durable and easy-to-clean floor, but very smooth surface. Floor surface will become slippery when exposed to any common contaminant. Floor must be kept clean and dry at all times, with extra vigilance during inclement weather. High slip potential.
M2 Steel sheet	No profile			≈	1–10	60/9 (stainless steel)	Highly durable and easy-to-clean floor but smooth surface. Floor surface will become slippery when exposed to any common contaminant. Rusting of steel will create rougher but inconsistent surface. High slip potential.
M3 Steel sheet	Durbar profile		~			80/70	Highly durable floor. Galvanised finish will lower slip potential, but high traffic wear will remove profiles, galvanised surface or applied paint, thereby increasing slip potential.
M4 GRP	Pultruded with no grit	—				75/60	Glass-reinforced plastics (GRPs) with extremely low slip potential. Can be used in internal and external environments and with highly viscous or semi-solid contaminant.
M5 GRP	Grating	—				75/60	As M12.
Resin							
Resin 1	Three-component solvent-free polyurethane screed	—			Off scale of surface roughness meter	63/47	Coarseness of aggregate too high to be measured by roughness metre and only true guide would be pendulum or ramp. Safe floor with low slip potential, but much harder to clean.

Flooring material	Finish	Slip potential in wet conditions			Typical roughness (value in μm)	Expected SRV, dry/wet	Comment
		Low	Moderate	High			
Resin 2	Two-component solvent-free epoxy safety system with clear 1.25 mm angular aggregate	–			Off scale of surface roughness meter	85/82	Coarseness of aggregate too high to be measured by roughness metre and only true guide would be pendulum or ramp. Safe floor with low slip potential, but much harder to clean.
Resin 3	Three-component solvent-free self-levelling screed			≈	1	85/13	Highly durable, easy-to-clean floor, but very smooth surface. Safe floor surface when dry and clean, but becomes very slippery when exposed to any common contaminant. High slip potential. Floor must be kept clean and dry at all times, with extra vigilance during inclement weather.
Resin 4	Two-component water-based epoxy coating with PVA flakes and clear polyurethane sealer		~		13	57/31	Durable floor, but with far more demanding cleaning regime than standard vinyl. Roughness and SRV creates much safer floor when exposed to water-based common contaminant. SRV will improve with traffic as aggregate exposed. Poor cleaning will increase slip potential.
Resin 5	Two-component solvent-free epoxy coating			≈	0.5	100/11	Highly durable, easy-to-clean floor, but very smooth surface. Safe floor surface when dry and clean, but becomes very slippery when exposed to any common containment. High slip potential. Floor must be kept clean and dry at all times, with extra vigilance during inclement weather.
Resin 6	Two-component solvent-free epoxy coating with 1.5 mm black granular aggregate	–			4	75/60	**Note** Nature of particles can confound roughness meter, leading to low readings. True figures only obtained from pendulum or ramp. Safe floor with low slip potential, but much harder to clean.
Resin 7	Two-component solvent-free epoxy coating with 1 mm black granular aggregate	–			7	66/53	As Resin 6.
Resin 8	Three-component solvent-free epoxy screed and two-component clear polyurethane sealer	–			22	60/43	As Resin 4.
Resin 9	Two-component water-based epoxy finish			≈	1	83/15	As Resin 5. High slip potential.
Resin 10	Two-component water-based epoxy finish with fine aggregate	–			16	58/50	As Resin 4.
Resin 11	Two-component solvent-free polyurethane coating			≈	1	93/10	As Resin 5. High slip potential.

Flooring material	Finish	Slip potential in wet conditions			Typical roughness (value in µm)	Expected SRV, dry/wet	Comment
		Low	Moderate	High			
Resin 12	Two-component solvent free polyurethane coating with 1.25 mm clear granular aggregate	–			6	73/59	**Note** Nature of particles can confound roughness meter, leading to low readings. True figures only obtained from pendulum or ramp.
Resin 13	Two-component solvent-free polyurethane coating with 1 mm black granular aggregate	–			6	70/53	**Note** Nature of particles can confound roughness meter, leading to low readings. True figures only obtained from pendulum or ramp.
Resin 14	Two-component water-based clear matt polyurethane lacquer		~		13	70/25	As Resin 4.
Resin 15	Two-component water-based epoxy finish	–			16	67/58	As Resin 4.
Textiles							
Tex1 Carpet	Tile	–			n/a	n/a	Carpets in tile or broadloom natural or man-made fibres do not usually create slipping hazards.
Tex2 Carpet	Broadloom	–			n/a	n/a	As Tex1.
Glass							
G1 Float glass	Untreated			≈	0.1	85/9	About as smooth a floor surface as exists. May wear rougher with traffic, but still likely to be unsuitably smooth. Slips almost guaranteed. Should not be used in foreseeably wet areas and inclined surfaces or externally under any circumstances. High slip potential.
G2 Float glass	Acid-etched or grit-blasted	–			25 – for grit-blasted	60/45	Etched or grit-blasted surface gives as rough a surface as required. Can be used for inclined surfaces but with care. Slip potential may be increased by wear and traffic creating smoother areas of floor. SRV should be verified before specifying and monitored during use.

KEY POINTS

◆ *The appropriate design of building entrances can significantly reduce the ingress of rainwater and dirt into the building, thereby reducing the risk of slipping.*

◆ *Building orientation, the provision of a canopy to shelter the entrance, appropriate matting, and the use of underfloor heating or hot air curtains will reduce water contamination of the indoor space.*

◆ *Stairs are frequently a location for serious slips and falls. The going size, surface material and type of nosing are all features that should be designed to minimise slip risk.*

◆ *Any pedestrian surface that is to be installed on a slope requires a higher slip resistance than if it was installed on the level.*

◆ *For external pedestrian spaces, a pedestrian surface with an SRV of 40 in the wet will have an acceptable level of slip risk, where there are no other risk factors present (slopes, vehicular access).*

◆ *The minimum SRV for a footbridge should be 45 (Highways Agency, 2004).*

◆ *A wet and dry slip resistance value of between 40 and 70 is generally safe for surfaces in stations.*

Courtesy Dave Richards, Arup

11.1 INTRODUCTION

This chapter considers the requirements of specific building elements and the requirements for each relating to minimising the risk of slipping. Requirements for entrances, general circulation spaces, stairs, ramps and slopes are presented before requirements for external spaces including footbridges and railway station platforms.

11.2 ENTRANCES

Given that most flooring has an adequate slip resistance value (SRV) in the dry, minimising the extent to which rainwater can come into a building is of critical importance in minimising slip risk. The following factors should be considered to minimise rain contamination at an entrance (Thorpe and Lemon, 2000):

- orientation and prevailing wind – by orientating the building away from the direction of the prevailing wind the risk of rain and other dirt and soil being blown through doors will be reduced

- canopy design – a well-designed canopy will shelter the outside of the door and reduce rainfall on the surface immediately outside the entrance. The more enclosed the canopy design, the more effective it will be in sheltering the entrance. Figure 11.2 illustrates where an opportunity has been missed to provide the entrance and the occupants with protection against the elements while giving a clear and bold indication of the entrance location

- positive internal air pressure – if the building has positive internal air pressure, when the entrance is opened warm air will tend to flow out, rather than wet and dirty air flowing in. Warm air in the entrance area will increase the rate of evaporation, reducing the time that the floor is wet and remains a higher slip risk

- underfloor heating – to speed up drying.

In addition, the type of door and the way it opens (ie if the prevailing wind tends to blow the door shut rather than open) can also reduce the ingress of water – see Figure 11.5. Other structures can also protect the door from prevailing wind – see Figure 3.3.

Figure 11.1   *Entrance with large canopy to reduce ingress of rain (Deborah Lazarus)*

Figure 11.2 *Entrance canopy serving no useful purpose in protecting the entrance*

Figure 11.3 *Building façade and canopy protecting entrance to building and also affording dry storage of trolleys under cover*

11.3 THRESHOLDS AND DOORS

A threshold presents the risk of both slipping and tripping.

To reduce the risk of slipping, the floor surface adjacent to an entrance should be chosen to ensure that it does not become slippery when wet (Roys, 2003). Alternatively, appropriate matting should be provided. The matting will only remain effective, however, if it is regularly cleaned and maintained as recommended by the manufacturer (BS 5395-1). Materials used at the threshold should be of a high enough SRV to counter any expected contaminant. Materials with very low SRVs, such as ceramic tiles, terrazzo and steel, are often inappropriately used to demarcate the threshold and areas surrounding the entrance matting system.

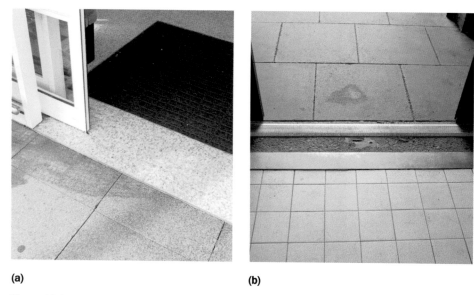

(a)   (b)

Figure 11.4   *Thresholds with low-SRV materials: (a) terrazzo tile; (b) stainless steel. The threshold in (b) could also pose a trip hazard*

Approved Documents M (ADM) and K (ADK) of the Building Regulations 2000 contain guidance on requirements for thresholds and doors. To minimise the risk of tripping, steps at the threshold should be avoided wherever possible. If not, the door should open over the higher level and the rise be no more than 150 mm.

Heavy doors that do not blow open in windy weather will reduce the occurrence of wind-driven rain or dust being blown into the building's entrance.

Figure 11.5   *Heavier doors installed to reduce occurrences of wind-driven rain or dust being blown into the building (Deborah Lazarus)*

ENTRANCE MATTING

Matting can be used to reduce the risk of slips at entrances by removing wet contamination from pedestrian footwear. It works best when used in conjunction with other controls including effective or enclosed canopies, heating and particularly underfloor heating and ventilation.

An effective matting system gives long-term protection to the building's interior from the ingress of dirt, mud, grit and moisture. It should have a high moisture retention, combined with ease of cleaning and maintenance; it should dry quickly and not show traffic tracks, being both heel- and wheel-proof; fungal and bacteria-resistant; and aesthetically acceptable.

BS 7953 identifies typical matting materials and provides some guidance on their selection, installation and maintenance. However, most currently available matting materials and systems appear to be primarily concerned with the removal of grit and dirt from shoes to protect the floor from damage.

Recently some materials that may be more effective at removing water from shoes have come to the market. Limited research suggests that natural fibre materials are the most effective, but further work is required.

Drained mat wells, which prevent the mats becoming waterlogged, are good practice in areas of high footfall such as underground stations.

It is difficult to be prescriptive about the number of footfalls required to remove moisture completely from footwear. The more footfalls the mat allows, the more likely it is to be effective. Some references state that the effective length of an entrance flooring system, or matting, should allow for a minimum of two footfalls for each foot (ie four paces), as this is considered sufficient to dry and clean the shoe (Thorpe and Lemon, 2000; BS 7953). If wheeled vehicles are expected to use the entrance, the matting should be long enough to allow for one complete rotation of the wheel (BS 7953). However, experience has shown that this is not usually sufficient. If wet footprints are visible beyond the matting then it is being compromised and further matting and/or other controls will have to be considered.

Soft entrance matting can be difficult to negotiate in a wheelchair, as it compresses under the wheels and inhibits the user's movement. It can also be a trip hazard to others using walking sticks or frames. To minimise this problem, BS 8300 states that dirt-collecting matting should be of dense construction.

Matting should extend to the threshold and should not leave a small strip of unprotected floor between threshold and matting – see Figure 11.4. It should be fixed securely such that it does not present a tripping hazard. It should be cleaned and maintained appropriately and regularly. If wet cleaning techniques are used then the mats must have sufficient time to dry before use. Worn matting may present a trip risk.

When supplementary matting is needed then it should be butted up to the fixed matting with no gaps and should not be allowed to curl or ruck so as to introduce a trip risk.

The positioning of matting should coincide with the way pedestrians actually use the entrance. The position and dimensions should prevent short-cut routes that avoid the matting.

Matting is functional and designers should avoid using it to make aesthetic design statements (Figure 11.6).

Figure 11.6 *Entrance matting that may not coincide with the direction in which pedestrians are likely to walk. It is also interspersed with low-slip-resistance tiles*

The pendulum test can be used to determine the slip resistance of mats. The property may be directional and this should be tested by the operator. Surface micro-roughness measurements cannot be made on the matting materials themselves but may provide additional useful information on metal, plastic, rubber or other inserts, often an integral part of the matting, used to help remove gross contamination from footwear.

11.5 GENERAL CIRCULATION SPACE

There is no explicit requirement for the slip resistance of internal floor areas beyond what is covered in the Health and Safety at Work Regulations, but the disabled access requirements in ADM state that in buildings other than dwellings the floor surface should be slip-resistant. BS 8300 reinforces this, stating that floor surfaces should be slip-resistant to give a firm foothold and good wheelgrip under all wet and dry conditions.

Figure 11.7 *Example of general circulation space: University of Limerick Foundation Building (courtesy Arup/Studioworks)*

The following common-sense requirements have been suggested (Roys, 2001) to minimise slipping accidents on all floors:

- make floors level and stable

- select a floor material that is unlikely to deteriorate, deform or detach in use

- secure floor coverings firmly to the floor

- avoid finishes that could become slippery in areas where they are likely to become wet or contaminated, such as kitchens, bathrooms or at entrances

- fix or back mats and rugs to avoid slips and trips

- choose flooring that is easy to clean and maintain

- changes in floor surface or inclination should be denoted by the use of colour contrast

- avoid, if possible, having floor finishes with very different slip characteristics adjacent to each other, eg carpet and polished marble.

11.6 STAIRS

Considerable effort has been spent in investigating the factors that exacerbate the likelihood of accidents on stairs. One in 10 of all accidents that occur in the home are on the stairs (Roys, 2001). The very young and elderly are particularly susceptible to falling accidents on stairs (BS 5395-1). Each year there are approximately 500 fatalities on domestic stairs and 100 fatalities on non-domestic stairs (Loo-Morrey, Hallas and Thorpe, 2004).

The severity of slips and falls during descent is greater than for ascent because the action of ascending means the subject will fall up only a few steps, but may fall down the entire flight during a fall in descent. Eighty per cent of injuries that occur on stairs happen during descent (Roys and Wright, 2003).

ADK (Reg K1) requires that "stairs, ladders and ramps shall be so designed, constructed and installed as to be safe for people moving between different levels in or about the building" . ADK contains the requirements for stairs, including the size of the treads, provision of handrails and headroom. Approved Document B (ADB) and BS 5588-5 are also relevant if the stairs are a means of escape from fire. ADM and BS 8300 cover the design of stairs for disabled access. ADM states that should there be a contradiction between Approved Documents K and M, the requirement in ADM takes precedence.

BS 8300 requires that in order to meet the requirements of disabled people the surface materials for internal steps and stairs should be as slip-resistant as possible, especially if they are likely to become wet as a result of their location or use or if spillage occurs.

Figure 11.8 *Steps with contrasting nosings to tread material, corduroy warning strip and three-level handrails, the very low one being for children (courtesy HSE)*

Factors in stair design of particular relevance to slips and trips are the size of the going, the slip resistance of the stair material and the use of nosings. These are addressed below. The risk of slipping increases if the tread and nosing are finished in a smooth material, if the steps are wet or if the edge of the step is rounded (Roys and Wright, 2003). Poor lighting may also increase the risk of slipping, as it reduces the ease with which the edge of the stair can be distinguished, particularly by users with poor eyesight, and hence can cause the user to misplace his or her foot. A lack of handrails increases the risk of falling, as the user cannot stabilise himself or herself while negotiating the stairs. In the case of the elderly, it is probable that there is not a single cause that causes a slip, but a combination of factors (Easterbrook *et al*, 2001), including handrail position and shape.

11.6.1 Going size

Figure 11.9 shows the appropriate terminology for describing the dimensions of stairs.

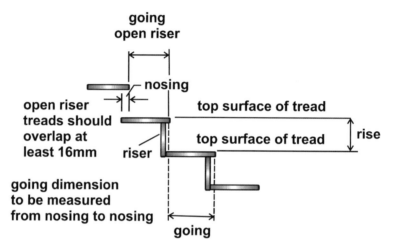

Figure 11.9   *Stair terminology (ADK)*

Going	The horizontal distance between the nosing of a tread and the nosing of the tread or landing next above it.
Nosing	The front edge of a tread.
Rise	The height between consecutive treads.
Tread	The upper surface of a step.

The most appropriate dimensions for stairs are determined by their location. They differ for stairs in private (in a single dwelling), institutional and assembly (in places where substantial numbers of people gather) and other (in all other buildings) uses. Institutional and assembly stairs may be used by many people moving relatively slowly, so the rise should be low and the going quite wide. In private stairs the going may be smaller and the rise higher because it is presumed that the occupant is familiar with the stairs. Public stairs are defined in BS 5395-1 as stairs where people are likely to move faster, so the rise can be higher. ADK contains requirements for the dimensions of the goings and rise of stairs as presented in Table 11.1.

According to ADK, regulation K1 can also be met by following the requirements of BS 5395-1. The latter states that the tread width should be sufficient to provide adequate support to the shod foot. It should allow at least part of the heel, when ascending the stair, to rest firmly on each step, and should permit descent without the need to place the foot at an awkward angle. As steps with a shallow rise tend to cause people to trip, this should be greater than 100 mm, and not exceed 220 mm for private stairs (see Table 11.1). According to BS 8300, the going size for domestic and non-domestic buildings should be 250–300 mm, with a preference for 300 mm, to improve ease of use for disabled users. This is larger than the ADK requirement. Where there is a conflict in stair dimensions, ADM takes precedence. Both BS 8300 and BS 5395-1 are about to be updated, and any confusion over the requirements should be resolved.

Table 11.1   *Maximum and minimum rise and going sizes (ADK; ADM; BS 5395-1; BS 8300)*

Source	ADK Max rise (mm)	ADK Min going (mm)	ADM Rise (mm)	ADM Going (mm)	BS 5395-1 Rise (mm)	BS 5395-1 Going (mm)	BS 8300 Rise (mm)	BS 8300 Going (mm)
Private stair	220†	220†	Refers to ADK requirements		100–220	225–350	150–170	250–300 (300 preferred for schools)
Institutional and assembly stair	180**	280*	–	–	100–180	280–350		
Other stair	190**	250*	–	–	–	–		
Public stair	–	–	–	–	100–190	250–350		
External steps and stairs other than dwellings	–	–	150–170	280–425	–	–		
Internal stairs other than dwellings (schools)	–	–	150–170	>250 (>280)	–	–		
Stairs up to a dwelling	–	–	75–150	>280	–	–		
Stairs within the entrance storey of private dwellings	–	–	Refers to ADK requirements		–	–		
Common stairs to private dwellings	–	–	<170	>250	–	–		

†   The maximum pitch for a private stair is 42°
*   If the area of a floor of the building is less than 100 m², the going may be reduced to 250 mm
**   For maximum rise for stairs providing the means of access for disabled people reference should be made to Approved Document M: Access and facilities for disabled people.

A study was undertaken (Thorpe and Loo-Morrey, 2003) using a variable going stair rig and 60 volunteers to determine the significance of going size on the safety and ease of use of stairs. Based on these results, it was recommended that the minimum going sizes be increased to above the critical going size as follows:

● private stairs = 250 mm

● public stairs = 300 mm

● assembly stairs = 350 mm (Thorpe and Loo-Morrey, 2003).

The dimensional consistency of a stair will also influence the risk of slipping or mis-stepping. It is unusual for the rise and going on any stair to be exact throughout the whole flight, even though the consistency is specified in ADK. Building tolerances with variations of 4–6 mm are common (Roys and Wright, 2003). Anything above these figures will lead to uncertainty on the part of the user as to where to place his or her feet and may result in an unanticipated overstep.

Figure 11.10 *Example of steps with uneven riser heights outside the tolerances allowed by the Building Regulations (ADK) (courtesy HSE)*

Figure 11.11 *Stair with different riser heights and materials (courtesy HSE)*

11.6.2   Slip resistance of stair material

The SRV of the stair tread material will influence the likelihood of slipping. If a contaminant is present the slip resistance will be reduced and finishes that present a low risk when dry may give a dramatically increased risk of slipping. When walking on the flat, floors with a SRV of greater than about 36 pose a low slip risk to users (1 in 1 000 000 users would experience difficulty in walking on such a level surface). Surfaces with an SRV of less than 25 have a high slip risk. However, when people walk up or down stairs, it is the toe which first makes contact with the walking surface, compared to the heel for walking on the flat. As a person's gait is different on a stair, the SRV requirements are unlikely to be the same. Research is not unanimous, but suggests that the friction requirement on the stairs may be appreciably lower than on a level surface (Beaumont *et al*, 2004).

The influence of stair material in dry, wet and glycerol-contaminated conditions on slip resistance was investigated (Thorpe and Loo-Morrey, 2003). The stairs were found to become more hazardous as the going size and SRV decreased. When the stairs were dry, no slipping occurred. In wet conditions, there were fewer problems for carpet, resilient safety floor and quarry tiles, which were the materials with a higher SRV. Difficulties in descent increased for glazed ceramic tiles in the wet as the going sizes decreased. The same, but less severe, results occurred for rubber flooring. On the whole, the SRV values of the stair surface did correctly rank the performance of the flooring materials used in the study, even though the critical SRV may differ on stairs from that on the level.

The study showed that the SRV required by users of stairs is lower than that required for walking on the level because the horizontal force between the foot and floor or step is lower. Users managed to negotiate the stairs under contaminated conditions with a SRV of 20 with little or no difficulty. Below this value of SRV, difficulties increased significantly, as might be expected; with larger goings, steps were still negotiable. As only two subjects participated in the test, further work would be necessary to confirm this.

As is the case with all flooring materials, the surface and nosing of a stair may become worn with time and the slip resistance will probably change.

The use of carpet reduces the likelihood of slipping, but it should be in good condition, secure and the nosings should be made clearly visible if the carpet is patterned (Roys, 2001). BS 8300 states that deep-pile carpet should not be used on stair treads.

Shiny polished surfaces should be avoided on stairs as they could cause glare, which might confuse the visually impaired. Shiny surfaces may also give the impression that the surface is slippery, even if this is not the case (BS 8300).

If the stair tread material is isotropic and its slip resistance is independent of direction, the pendulum can be used to measure the slip resistance of individual treads, provided the stair is wide enough to accommodate the full swing of the pendulum. The pendulum tester has to be set up so that is supported along the front edge from the tread below, or restrained by ad hoc bracing from the above. Surface roughness measurement should be used to determine whether the stir tread is isotropic or not (UKSRG, 2005).

11.6.3 Proprietary nosings

A proprietary nosing is a strip of material that is applied over the front of a step. They are available in a wide range of materials and grades. Proprietary nosings are installed to adjust the user's interaction with the stairs in the following ways:

- to improve the slip characteristic of the stair
- to increase colour contrast between the going and the nose of the stair to help users place their feet more accurately and avoid overstepping
- to protect the surface of the step from wear.

The main types of nosing are:

- silicon carbide
- tapes
- synthetic rubber
- metal
- PVC
- aluminium with PVC inserts.

The material used on the surface of the tread will influence the decision on whether to install a nosing or not. Carpet, if installed on heavily used stairs, such as in an office or public building, can become worn smooth at the edge of each stair. The use of a nosing would protect the carpet against becoming smooth and wearing through, thereby reducing the exposure of users to slip and trip hazards. Rough materials, such as clay pavers, clay tiles with a carborundum finish, cork tiles, mastic asphalt and resin-based tiles with enhanced slip resistance normally provide sufficient slip resistance on stairs, even in wet conditions, without the need for proprietary nosings. In some situations where the going is small, or if there is significant contamination, the addition of a nosing may reduce slip risk. On smooth materials that have an inherently lower slip resistance, such as glazed ceramic tiles, float glass, PVC, rubber, stainless steel, polished terrazzo and finished timber, adding an appropriate nosing, with a slip-resistant finish in the correct place, could reduce the likelihood of slips (Roys and Wright, 2003)

Figure 11.12 *Terrazzo stairs with slip-resistant nosing correctly placed at the front of each stair tread*

Square-shaped nosings are generally thought to be the most effective in reducing slips. A nosing can be used to change the shape of the edge of the stair and to increase the effective size of the tread. For example, if the stair edge is rounded, a nosing with a square profile would increase the going size slightly. Nosing manufacturers advise against this, as they consider that the same profile should be added in preference to one with a different profile, since the fixing may be less effective and therefore less safe. Although proprietary nosings that overhang the edge of the stair increase the size of the effective tread depth, they can cause difficulty for ambulant disabled users ascending the stair. Alternative designs are available that overhang the edge of the stair but taper gradually back to the riser, thereby reducing the risk of tripping. However, the selection of the best-shaped nosing is not always clear-cut. A nosing that increases the going size is likely to reduce the incidence of slips, but, should a slip occur, a sharp-profiled nosing would cause more damage to the victim than one with a rounded profile. In the absence of any colour contrast at the edge of the step a slightly rounded nosing profile may be beneficial because it will reflect the light in different directions, thereby increasing the definition of the edge of the stair (Roys and Wright, 2003).

If proprietary nosings are installed, they should be consistent for the entire flight of stairs and flush with the surface of the tread. It is essential that they be securely fixed and unable to move when stepped on. A fatality occurred when a square-edge nosing partially fixed over an existing bullnosed timber stair tread became completely detached during use, throwing the person forward and down the stairs and against a corridor wall.

Approved Document M-compliant nosings will be required in most buildings. These incorporate a 55 mm visible contrasting strip on both the riser and the going.

BS 5395-1 recommends that the surface roughness of a nosing material be at least 20 microns.

In-situ pendulum testing of stair nosings perpendicular to the stair is impractical because the pendulum is prevented from swinging by the step above. To undertake an assessment of a stair nosing, it would need to be removed and made into a flat tile made up of several lengths placed adjacent to each other (UKSRG, 2005).

Research was conducted by BRE and HSL (Thorpe and Loo-Morrey, 2003) into the effects of nosings on dry, uncontaminated stairs. The research found that, to be effective in reducing slips, nosings should:

- provide roughness at the very edge of the tread
- for nosings that curve around the front of the tread the roughness must be effective over the entire radius of the nosing
- increase the effective tread depth
- indicate the edge of the step by a clear colour contrast.

This study did not look at the influence of nosings once they become worn, and it is possible that their effectiveness will change with time.

11.7 VERTICAL ACCESS LADDERS

ADK states that stairs, ladders and walkways in industrial buildings should be constructed according to BS 5395-3 (for industrial stairs) or BS 4211 (for ladders). For industrial stairs, BS 5395-3 states that treads should be slip-resistant or at least have a slip-resistant nosing not less than 25 mm wide. Treads on open riser stairs should overlap not less than 16 mm and have a nosing depth in the range 25–50 mm to aid visibility. For ladders (type A or B), BS 4211 states that work platforms, where provided, shall be at least 2.0 m × 1.0 m on plan and have a slip-resistant finish.

Figure 11.13 *Access ladder complying with BS 4211 (courtesy Guardrail Group)*

The BS EN ISO 14122 standard series addresses the topic of permanent means of access to machinery in relation to safety. It comprises four parts:

- Part 1: *Choice of a fixed means of access between two levels*
- Part 2: *Working platforms and walkways*
- Part 3: *Stairs, step ladders and guard-rails*
- Part 4: *Fixed ladders*.

It is anticipated that the approach taken in the BS EN ISO 14122 series will be incorporated into future standards and approved documents. It requires all surfaces – horizontal, inclined or vertical – to have enhanced slip resistance.

There have been several deaths in the United Kingdom caused by people falling from these structures, particularly from the ladder rungs down and over the guardrails. It is thought that the initiator has been a loss of three-point contact with the ladder because the person slipped off the rungs.

11.8 RAMPS AND SLOPES

Ramps are beneficial for wheelchair users (and their carers) and people pushing roll cages in supermarkets, warehouses, laundries and hotels as well as people pushing prams, pushchairs and bicycles. Gradients should be as shallow as practicable, as steep gradients create difficulties for those wheelchair users who lack the strength to propel themselves up a slope or have difficulty in slowing down or stopping when descending. However, there may be circumstances, in shop fit-outs for example, where a steeper gradient than the maximum shown in Table 11.2 may be necessary over a short distance. The case for such a solution should be made in the access statement (ADM).

ADM and BS 8300 specify requirements for ramps in non-domestic buildings, including the need for a ramped approach to have a surface that reduces the risk of slipping when dry or wet. It should also be of a colour that contrasts visually with that of the landings to aid the visually impaired.

Table 11.2 *Recommended maximum gradients for ramps and additional coefficient of friction and SRV requirements on ramps (ADM; UKSRG, 2000)*

Going of a flight	Maximum gradient	Maximum rise	Additional coefficient of friction requirement	Approximate SRV for low slip risk
	On the level	0 mm	0	36.0
10 m	1:20	500 mm	0.05	41.0
5 m	1:15	333 mm	0.067	42.7
2 m	1:12	166 mm	0.083	44.3

Figure 11.14 *Example of a ramp that is not marked in a contrasting colour and that is made more difficult to distinguish because of the patterned carpet (courtesy HSE)*

Adverse weather conditions increase the risk of slipping on an external ramp. ADM therefore states that it is beneficial to have steps as well as a ramp.

The SRV of the surfaces adjacent to the stair should be similar to that of the ramp to minimise risk of stumbling. The surface of both should be slip-resistant (ADM).

Any pedestrian surface that is to be installed on a slope requires a higher slip resistance than if it was installed on the level. To estimate the coefficient of friction required on a slope, the tangent of the angle of the slope should be added to the requirement for the level (see Figure 11.15 and Table 11.1). LUL requires the sloped surfaces of all publicly accessible ramps to have an SRV of at least 65 (LUL, 2005b). It does not state whether this is wet or dry, although it should be appropriate for the anticipated service conditions.

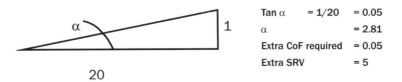

Tan α	= 1/20	= 0.05
α		= 2.81
Extra CoF required		= 0.05
Extra SRV		= 5

Figure 11.15 *Worked example of additional SRV required for a 1:20 gradient slope surface*

Ramped walkway in station

A ramp was surfaced with ceramic tiles that became slippery when wet, resulting in several slip accidents.

Initially, railway station management arranged for the ramps to be treated with an acid etch in order to increase the slip resistance of the tiles. This did help to reduce the number of slipping accidents. However, it was felt that more could be done. Later, a second anti-slip surface was applied. This was a reinforced plastic covering that was simply glued on top of the old ceramic tiled surface, giving the visual effect of a brick paver surface.

In 12 months slipping accidents reduced by 50 per cent and it is projected that the payback period for the £35 000 cost of the work will be less than two years based on reduced claims alone.

Proposals are being formulated to extend the treatment to the passenger concourse and other areas at the station.

Source: HSE, <www.hse.gov.uk/slips/experience/railway.htm>, accessed 1 Dec 2005

HSL research into methods of testing sloped surfaces indicates that the pendulum test is the most accurate methodology available. However, it only assesses the material's slipperiness, and not the extra risk of slipping as a result of the incline (Lemon *et al*, 2002b). Any surface finish that reduces the slip resistance should not be used on a ramp.

Cox and O'Sullivan (1995) state that evidence suggests that ramps may be a potential source of accidents, particularly in buildings where crowds gather.

Deep-pile carpets should not be used on internal ramps, as its resistance increases the effort needed to propel a wheelchair (BS 8300).

Figure 11.16   *Example of an access ramp with contrasting sloped surface at the rear entrance to a church in the City of London (courtesy Rupert Perkins)*

11.9   EXTERNAL PEDESTRIAN AREAS

This covers a wide range of external spaces, from pavements and paths to general paved areas.

The Department for Transport has produced guidance on the design of a wide range of public spaces, with a view to ensuring that they are easily negotiable by disabled people (DfT, 2002). Its guidance for footways (ie pavements adjacent to highways), footpaths and pedestrian areas states that they should be firm, slip-resistant in wet and dry conditions, and should have a dry SRV of between 35 and 45. This guidance is confusing, however, as a surface with a dry SRV of 35 would have a lower SRV in wet conditions. For the surface to have adequate slip resistance in the wet, the wet value should exceed 35 and the dry value would be higher. Furthermore, they should not be made of reflective material.

To minimise the risk of slipping on external spaces the following actions should be taken:

- where possible, level access to the building should be provided
- external surfaces should be in good repair, unobstructed and well drained to reduce the risk of slipping in wet or icy conditions (Roys, 2001)
- leaves, mud and algal growth should be cleaned from external pedestrian surfaces
- effective procedures to deal with snow or ice should be instated (HSE, 2003b).

The DfT (2002) guidance also includes recommendations for the provision of gates and fences, design of steps – hand-rails, lighting, going dimensions etc – which are beyond the scope of this publication.

ADM provides guidance on the surface of disabled parking spaces, which should be firm, durable and slip-resistant, with undulations not exceeding 3 mm under a 1 m straight-edge for formless materials. Inappropriate materials include loose sand and gravel (ADM).

An SRV of 36 or more, as measured in the wet condition by a TRL pendulum tester, is considered to be sufficient and not present an appreciable risk of slipping to pedestrians, where there are no additional risk factors, such as vehicular traffic or significant gradients. Such a surface has a low potential for slip, when tested using either type of slider (Four-S rubber or TRRL rubber). With additional risk factors, a value of 45 is suggested (County Surveyors' Society, 1996).

For areas extensively used by horse riders, the normal criteria for vehicular traffic are appropriate, but the use of surface texture on inclines is suggested (County Surveyors' Society, 1996).

Figure 11.17   *Example of unacceptable uneven paving (courtesy HSE)*

Figure 11.18   *Pedestrian circulation space outside railway station*

11.10   FOOTBRIDGES

The slip-resistance requirements for footbridges are covered in Part 8 of the *Design manual for roads and bridges* (HA, 2004). It has two key requirements for slip resistance:

The minimum slip resistance of traversed areas shall be equivalent to a mean corrected Pendulum Test Value of 45 units using a standard skid resistance pendulum test (BS EN 13036-4). This requires a "Four-S" or CEN rubber slider on a wet test surface

Cover plates to gaps and joints shall be set flush with the top of the surfacing to prevent tripping, and the upper surfaces shall be suitably profiled or treated to reduce the likelihood of slippage. The upper face of cover plates to expansion joints at deck level shall be provided with a suitable slip resistant coating

Figure 11.19   *Timber footbridge with slip-resistant strips, Corporation Street, Manchester. For external use the strips should be spaced so that every footfall lands on at least one strip (courtesy Bernard O'Sullivan (Inside Out))*

11.11   RAILWAYS

Slips, trips, falls and related accidents, eg on stairs, are the most prevalent cause of harm to passengers at stations on the national railway network in Great Britain. In 2004, nearly 50 per cent of accidental harm at stations resulted from such accidents, including three fatalities, more than 125 major injuries, and nearly 2800 minor injuries. Most of these accidents are typical of those found in other sectors, and the opportunity to learn from other industries is being researched.

Figure 11.20 shows the total harm to passengers and members of the public at stations from slips, trips and falls, grouped by type of accident. It shows that:

- most harm does not occur in the form of fatalities, although on average one fatality a year occurs on stairs or escalators. There were two fatalities on stairs in 2004. Stairs are a significant hazard for those suffering incapacity, infirmity or the influence of alcohol

- fatalities occurring as a result of accidents on level ground are exceptional. In 2004, however, a person fell while running for a train and died two days later from the head injuries incurred, despite having got up and caught the train

- minor injuries contribute a large proportion of the harm for almost all categories.

While overall harm from slips, trips and falls shows no improvement at stations, passengers have progressively made more journeys, and large stations are increasingly used as shopping centres. The level of usage of stations has increased significantly. Network Rail estimates that the usage of its managed stations has increased by more than the increase in journeys. Maintaining constant numbers of accidents therefore represents progressively improving safety per usage (RSSB, 2005).

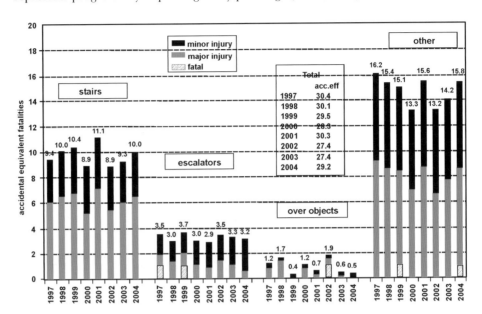

Figure 11.20 *Slips, trips, falls and related accidents at stations (RSSB, 2005).*

Acc. ef = accidental equivalent fatalities. One fatality is equivalent to 10 major injuries (eg broken ankle) and 200 minor ones (eg sprained ankle)

Department for Transport guidance (2002) for inclusive mobility requires platforms to be even, slip-resistant and non-reflective.

The Strategic Rail Authority recently published a revised code of practice (SRA, 2005) on train and station services for disabled passengers. This code should be used as the main reference document for disability provision in the railway environment.

The SRA code of practice covers all aspects of station design and states that all floors should "have some slip resistance when wet or dry" (SRA, 2005). An SRV of between 40 and 70 is generally considered safe. This must be above 50 for platforms in the open and may be as low as 40 where they are fully covered. The test must be conducted in both wet and dry conditions, and measured with four-S rubber, on a pendulum test. Surfaces with values outside this range are likely to be slippery or too rough and therefore more likely to contribute to accidents. If necessary, existing floor surfaces should be treated to improve their slip resistance. Where two materials abut each other they should have a similar level of slip resistance, otherwise the foot, walking frame or wheel will be abruptly slowed or caused to slip.

New hard floor surfaces, such as ceramic tiles, natural stone, concrete or terrazzo, which are widely used in commercial environments for their durability, should use an additive, such as carborundum, to make them more slip-resistant. Mats can be used to reduce risks, as long as they are fixed and flush with the floor, so there is no danger of tripping over them (SRA, 2005); see Section 11.4.

In buildings that are open to the elements, such as most railway and bus stations, contaminants such as pigeon droppings can add considerably to the slipperiness of surfaces, especially when subject to wetting by rain etc. They can be minimised (with varying degrees of success) using the following:

- for new construction, design to remove or box-in ledges, overhead girders or other opportunities for roosting

- netting or making roosting difficult on cross-beams and other surfaces, for examples with spikes

- gels are available, but these are not very effective and can be harmful to birds

- real or dummy birds of prey have been found to have some benefits, but research on this approach is inconclusive

- ensuring passengers etc do not feed the birds.

In addition to the above measures, the cleaning regime must take account of the need to remove deposits as they occur.

(a)   **(b)**   **(c)**

Figure 11.21   *Problems with pigeons: (a) nesting on horizontal structure (courtesy Michael Woods); (b) the resulting slip hazard below (courtesy Michael Woods); (c) mesh installed to exclude pigeons from possible roosting locations (courtesy S G Spark)*

Figure 11.22   *Example of a tactile warning surface near the platform edge (courtesy HSL)*

11.12 GRATINGS

Gratings are widely used in industrial situations where contamination of the floor is likely to occur. They are also used in stairs and walkways. They are typically made from steel, stainless steel, aluminium or GRP, and are designed according to BS 4592-1 or BS 4592-2. Where they are associated with machinery, they are designed to BS EN ISO 14122 -1, -2, -3. BS EN ISO 14122-2 states that walkways and working platforms shall be designed and built in such a way that the walking surfaces have durable slip-resistant properties.

The use of an anti-slip grit-based coating will increase the slip resistance of many gratings. Anti-slip nosings or treads should be used on all access stairs of this kind.

London Underground require the "Heelguard" type of grating to be used.

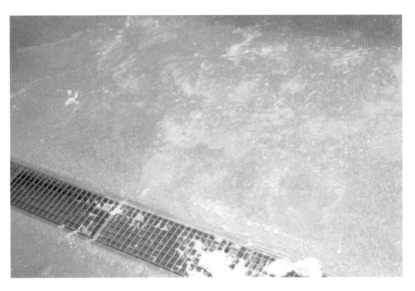

Figure 11.23 *Grating over gulley in an industrial floor (courtesy HSL)*

12 Case studies

Courtesy HSE

12.1 GENERAL

This chapter provides a variety of examples of slip incidents, demonstrating both the underlying causes of slips and, in some cases, their consequences.

The examples are drawn from various sources; they include case studies provided by the HSE and details of forensic investigations carried out by the HSL.

While slip incidents may lead directly to injury or discomfort from impact with the ground or, for example, furniture or machinery, there are also numerous cases recorded where the initial slip has then led to much more serious injury or even fatality. Under the current system of reporting and classification system, set out in the Reporting of Injuries, Diseases and Dangerous Occurrences Regulations 1985 and 1995 (RIDDOR), no differentiation is made between the initial factors causing a pedestrian to fall and impact with the underfoot surface due to loss of balance. All such accidents are classified as "slips, trips and falls on a level" (STFL). It is the cause of the injury and not the initiating event, which may have been a slip or trip, which is generally recorded, which leads to a significant under-reporting of STFL accidents.

> **Slip on a steel plate**
>
> A cleaner in a hotel had to empty rubbish from small bins into large plastic bags in his work area. When the bags were full, he carried them to a skip near the loading dock. To complete this task he had to cross an area of the dock, which was exposed to the weather.
>
> The flooring of the loading dock consisted of square steel plates. These plates used to have small rectangular indentations to provide a degree of slip resistance. However, the plates had been worn smooth by trolleys and pedestrian traffic. While carrying a full bag in each hand, the cleaner slipped and fell on the wet surface. As he fell, he hit his head on a steel door-frame, causing lacerations and neck injury, which resulted in a chronic disability.
>
> Source: *Preventing slips, trips and falls – guidance note* (WorkCover, 1998)

12.2   HSE CASE STUDIES

12.2.1   *Catering*

The HSE Slips and Trips website reports that some slipping accidents within the catering industry have resulted in burn injuries following contact with open frying ranges. A slip in the kitchen while carrying a knife has resulted in a fatality, as illustrated below. Chapter 1 and Appendix 3 also refer to the often unrecorded knock-on effects of slips.

> **Death from kitchen slip or trip**
>
> A mother of two died in a tragic accident recently as she worked alongside her daughter in a residential care home. The 51-year-old woman slipped or tripped on the kitchen floor and a large knife she was carrying severed an artery in her neck.
>
> Derwentside District Council investigated the incident assisted by technical experts from the Health and Safety Executive and the Health and Safety Laboratory.
>
> Source: HSE, <www.hse.gov.uk/slips/issues.htm>, accessed 1 Dec 2005

Fractured skull from a "wholly preventable" catering slip

- A kitchen worker fractures her skull.
- Injured member of staff unlikely to work again.
- Prosecuted employer ordered to pay more than £36 000 including prosecution costs.

On hearing a prosecution case against a retailer arising from a slip accident in a store restaurant the district judge said:

> This was a very serious accident, one which was wholly preventable. Every employer has a duty under law to protect its employees from physical harm – something that [the company] blatantly failed to do. There had been four similar accidents in their kitchen during the previous 12 months, yet they still failed to act.

When a local authority health and safety inspector visited the restaurant to investigate the circumstances surrounding the serious injury sustained by a member of staff he quickly became concerned that the floor surface in the kitchen was very slippery with even the smallest amounts of water or grease on it.

Partial matting and bare kitchen floor tiles

The inspector found that the tiled floor appeared to be in good condition yet it still felt slippery even when the tiles looked to be clean and dry. When just small amounts of water got on to the floor it was found to be very slippery. To make matters worse, some areas of the kitchen floor sloped, increasing the slip risk. Kitchen staff could be seen walking with a very peculiar gait to try to avoid slipping. Floor "safety" mats had been put down in some parts of the kitchen, such as in the dish-wash area, but these were slippery to walk on too, especially when wet. Cleaners had removed these mats at the time of the worker's accident, leaving her to walk on the slippery tiled floor, which quickly became contaminated with food waste, water and oily residues.

The worker experienced an uncontrollable slip and hit her head on the hard tiled floor. She was rushed to hospital where she drifted in and out of consciousness, suffered seizures and spent a lengthy period in the hospital's high-dependency unit.

The safety matting

The company's safety records examined by the inspector showed that there had been other slip incidents in the area but the response had been to provide the most heavily contaminated areas, such as the dish-wash, with matting – matting that was itself slippery, especially when wet. Scientists from the Health and Safety Laboratory examined the tiled floor surface and the matting and confirmed that the inspector was right to conclude that risk of slipping was unacceptably high because of the lack of slip resistance of the tiles and the matting. The floor surface was clearly not "fit for purpose".

The company was also aware of numerous other slip accidents in similar situations at their other sites around the country. The inspector served an improvement notice on the company to require it to deal with the slip risks to employees. The company considered various options, but after concluding they were inadequate or inappropriate, it eventually replaced the floor surface with one that was suitable for use in an area where the total elimination of floor contaminants would never be possible. The new floor was specified to provide enough grip, even in wet or contaminated conditions.

Source: HSE, <hse.gov.uk/slips/experience/skullfracture.htm>, accessed 1 Dec 2005

The case study below gives details of the devastating consequences in terms of human cost of a particular set of slip incidents within the healthcare sector.

Slip incidents at a hospital resulting in amputation

Alison Hockaday worked in a hospital as an occupational therapy assistant. Her first injury happened in 1986 when she slipped on wet leaves on the entrance steps to the hospital, badly twisting her knee. In 1990 she slipped on a wet vinyl floor in the hospital, fracturing her right ankle. Alison continued to suffer considerable pain and disability in both knee and ankle, requiring numerous operations and eventually leading to a below-knee amputation of her right leg. She can only wear her prosthetic limb for a short time indoors and has to use a wheelchair outside. Alison has not worked since 1992, when her employment was terminated on the grounds of ill health. She received £600 000 compensation, but can no longer take part in the sports she used to enjoy regularly.

Alison's own story

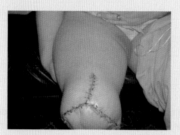

I worked as a technical instructor in a large hospital. On 16 November 1986 I was entering the main hospital whilst taking post to the sorting office when I slipped on decomposing leaves on the concrete steps. Fresh leaves had also fallen on top indicating the decomposing leaves had been there for some time. I fell heavily onto my right knee. I went home and returned to work the next day, but my knee was extremely swollen, painful and still bleeding. I was sent to hospital, where it was sutured and x-rayed, and it was found that I had damaged my kneecap, but hopefully time would heal it. I was still in pain three months later and I was told that I needed surgery. The leaves were supposed to be cleared on a daily basis, but due to staff shortages this was not done.

On 2 March 1992 I slipped again, this time on an unmarked wet floor. Snow was falling outside when I entered the building to start work. I went to my office changed out of my boots and into my flat shoes. I left my office to do attendance figures, but unknown to me someone had mopped the floor and failed to dry it or place any barriers or warning signs out around the area. In addition, the door mat, which was used daily by approximately 100 staff and patients, was too small for the area of tiled floor and it did not have the capacity to absorb all of the water being brought in by pedestrians. Some of the water was being transferred on to the dry floor.

I slipped and fell directly on to my right ankle. I was assisted to a chair and it was evident that the ankle was broken; it was also very contorted. I was taken to hospital and placed in plaster for six weeks. However, the plaster was taken off after three weeks, as the foot remained contorted. Over the next few years I faced some 32 operations to try to save the foot, but eventually I was told that I had dystonia caused by the accident. Amputation was the only solution, as my toes had by now lost their feeling and were turning black. I had my leg amputated on 30 May 1997.

One week after my fall the floors in the unit were made non-slip.

As anyone can imagine, these easily avoidable accidents have had a horrendous effect on my life. I was an active 21-year-old, only just married when I had my first accident. I remained working until my second accident and enjoyed dancing, aerobics and jogging, but I cannot do any of these activities now and spend a lot of my time in a wheelchair.

Although I have received compensation, I have lost my job and I will never work again due to ongoing problems. My husband has also lost his business because he now has to care for me. No amount of money compensates or prepares you for what has happened to me.

Source: HSE, <www.hse.gov.uk/slips/experience/leg.htm>, accessed 1 Dec 2005

The Cumberland Infirmary, Carlisle

The Cumberland Infirmary in Carlisle is a Private Finance Initiative hospital that was opened in the late 1990s. It was designed in accordance with the agreed NHS Estates *Specification manual*.

The main entrance comprises a central revolving door flanked by two automatically opening doors for wheelchair users, disabled people and those carrying stretchers. A canopy extends across the full width of the entrance, but it is not particularly effective at shielding the entrance from the elements.

This is an area where the user profile goes across all spectrums of age, gender, ability and gait – a profile that is common for public areas but which here includes a disproportionate number of disabled users.

Entrance door

Immediately inside the doors there is a reasonable matting system, then several metres of dark grey polished slate followed by terrazzo tiles.

Both tiles have a surface roughness of around 5 micron and are therefore very slippery when wet. There is also high reflection of sunlight on both tile surfaces, producing glare and causing further problems for the user.

Entrance matting

There is no control over footwear except where staff are involved. Even here the opportunity to exercise some degree of control is lost, as NHS policy on footwear is that it should be black, comfortable and "sensible", but is not required either explicitly or implicitly to be effective in preventing slips.

A main "street" runs through the building from the entrance. High levels of glazing produce high reflection beyond the entrance area – and also lead to thermal gain, with a very noticeably high temperature. The café area has the same flooring as the rest of the street; when the image below was taken, liquid spillages and food detritus were noticeable on the floor.

Polished slate floor

Continued overleaf

The Cumberland Infirmary, Carlisle (*contd*)

An assessment of the floor was made using the SAT, with the results putting the floor in the "significant slip risk" category.

Roughness readings: 7.5, 2.4, 3.2, 4.7, 6.5, 5.5, 3.3, 4.8, 6.1, 7.5.

- Water-based contamination
- Medium levels during inclement weather
- Cleaned regularly through the day
- Rotary buffer and polisher through the day
- Soon recontaminated
- No control over shoes
- 24-hour multi-users – staff, visitors, old, young, injured and disabled
- Pushing and pulling lifting and carrying
- High glare during sunny weather.

SAT assessment 34 – significant slip risk

Hospital management is aware that the area will be a problem in wet weather, but it had no control over selection of finishes at design stage and now has to control the problem by extra management and more frequent cleaning during wet weather. Vigilant staff man the first reception point (around 20 metres from doors) with a brief to make any necessary reports.

This shows use of the SAT to assess an existing floor and the provision of an appropriate management regime based on the results obtained. It also demonstrates how more careful consideration of the flooring specification initially, together with management involvement at an early stage, could have produced a more effective and safer solution from the outset.

Source: HSE

Healthcare sector accident statistics

Slips and trips are the main types of accident affecting workers and patients in the healthcare sector. In a TUC survey, 44 per cent of the total slips and trips occurred in the health services, compared with only 16 per cent in public administration. While the number of slips and trips reported on average for health service workplaces (76) partly reflects the size of hospital workforces, it also suggests a major problem compared with public administration (16–20) where workforces are just as large.

In addition, simple slip and trip injuries such as a broken bone often lead to potentially fatal complications such as thrombosis (blood clots) or embolisms (blood vessels becoming blocked) in older people.

Source: Unison (2004)

Fast food outlets

Wet mopping can increase the risk of slipping

HSE researchers were in a fast food outlet observing various activities in connection with slips and trips.

A customer bought food and drink, but spilled some coffee on the way to sit down. The spillage was small, about the size of a 50 pence piece.

The fast food company was aware of the risk of slipping from liquids or food spilled on their smooth floors and, almost immediately, a member of staff came to deal with the problem. They mopped up the spillage (and also the surrounding area because it looked dirty), squeezed out the mop and went over the whole area again, leaving an area of approximately 2 m² mop-dry.

The researchers took measurements on the mop-dry area using pendulum and surface micro-roughness techniques, and also timed how long it took to dry completely.

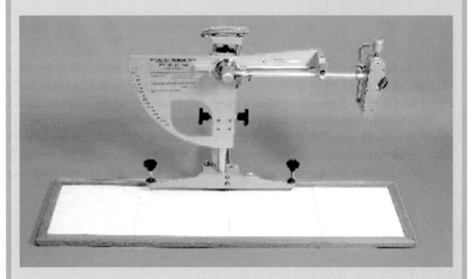

Taking measurements using the pendulum

The area of mopped floor, now almost indistinguishable in appearance from the rest of the floor, took approximately seven minutes to dry, and measurements showed that the area was extremely slippery during that time.

Research has shown that many slips are caused by a sudden change in floor surface characteristics. In this case, for the seven minutes until the floor dried completely, it would be difficult for customers and staff to realise they were walking from a safe to an unsafe surface.

The company, which had a good awareness of slips and trips risks and an efficient system for identifying spillages, had actually increased the risk of slipping because of the method of cleaning used in this instance. Simply cleaning up this spillage, and drying the small area of floor with a paper towel, would have been far better.

The fast food company is considering making a change to its spillage and general cleaning procedures.

Source: HSE, <www.hse.gov.uk/slips/experience/wet-mopping.html>, accessed 1 Dec 2005

Fast food restaurant owners ponder options

Which are the right slip prevention options for us?

A local authority environmental health officer (EHO) visited a fast-food restaurant (one of a chain) to investigate a slip accident in which the arm of a female employee was broken. The EHO identified several good aspects to the safety standards on site – a generally positive company attitude to safety, adequate training, well-kept documentation and records, proper floor cleaning systems. However, the busy servery area was found to be very slippery. The nature of the business meant that the floor in this area was bound to become wet at some stage. When the EHO spoke to members of staff they stated that the incident "had been waiting to happen".

Investigation revealed that several floor areas had been replaced in the past. However, the floor covering in the servery seemed to be the original, having been in place for more than 17 years. The EHO looked at the accident records and found that more slip incidents had taken place in the servery than elsewhere. Discussion with the duty manager revealed that the company response to this had been to deploy "Caution! Slippery floor" signs.

The owners arranged for testing to be carried out on the floors to find out about the surface roughness. This was found to quite good in most parts of the premises, but not in the servery. The original tiling in the servery had very low surface roughness – much less than was needed to be able to provide grip in a kitchen or servery. The EHO used this floor roughness information with the Slips Assessment Tool software (then undergoing field trials), which takes into account information about work activities, the environment, likely spillages and other relevant factors. The assessment indicated that there was a significant risk of slip injury – an indication borne out by the area's incident history.

The company was advised that the floor surface in the servery was at the heart of the problem and that the floor should be first thing to be looked at. The owners, however, wanted to try some special "anti-slip" overshoes for their staff. Despite the EHO's advice that the working environment should be put right before considering the use of personal protective equipment, the overshoe trial went ahead at three of the owners' sites.

Experience showed that the mainly young staff were reluctant to wear the overshoes (a fashion issue!). Enforcing the wearing of the overshoes impaired good staff relations, a good fit was hard to achieve as the only sizes available for the trial were "small", "medium" and "large", and tripping problems increased. Although the overshoes did provide extra slip resistance (so might be a viable option in some circumstances) the associated problems at the site negated the benefit.

The owners decided that the best option was, as the EHO had originally suggested, to tackle the problem in the work environment. The relatively small cost of reflooring (extended by the company to areas beyond the highest-risk servery) led to the conclusion that it would probably have been cheaper in staff and administration time to have pursued this option much earlier. Technical specifications of the proposed floor covering were obtained to ensure that it was suitable for the purpose – providing good slip resistance and being readily cleanable to meet food hygiene requirements.

A subsequent review showed that the slip accident rate had reduced by 70 per cent since the floor covering had been replaced and staff on site were much happier with the conditions.

Source: HSE, <www.hse.gov.uk/slips/experience/fastfood.htm>, accessed 1 Dec 2005

Public transport interchanges

Public transport interchanges are areas with a high risk of slip incidents, because:

- they are frequently crowded
- users are often in a hurry
- users may be encumbered
- users may be distracted (looking for journey details etc)
- spillages of beverages and food are common
- there is a high risk of litter and other contaminants that may adversely affect the slip resistance of the flooring
- wet contamination from precipitation occurs, with water brought in on footwear, umbrellas, luggage etc
- dirt is brought in from a variety of sources. When the conditions are wet as well, this has the potential to increase the slipperiness significantly.

Only some of the elements listed above are controllable. At railway stations on wet days yellow cones warning of slippery surfaces are a common sight, but more permanent solutions are possible. There is also the problem, noted in earlier chapters, of learned irrelevance, where the signs are left in position semi-permanently and are thus ignored, even in conditions where their message is valid.

Figure 12.1   *Warning cone permanently located at railway station (courtesy Deborah Lazarus)*

Using the recommended hierarchy of control levels, the areas where action can be taken by the operating company are:

- reducing floor surface contamination to negligible levels by regular and rigorous cleaning and drying
- modifying the floor surface
- replacing the floor surface where this is indicated.

Terrazzo flooring on station concourse

A railway company was concerned at the number of slipping and falling accidents on the terrazzo floor of a station concourse.

It engaged contractors to undertake a proprietary treatment process. This involved diamond grinding to produce a perfectly flat surface, followed by the application of a protective and slip-resistant coating.

The treatment and a modified cleaning regime has resulted in a surface that has a greatly enhanced appearance and is subjectively and demonstrably more slip-resistant. In addition, in the 18 months following treatment no civil claims arose from slips on the concourse compared with an average, previously, of three or four per year.

The railway company was so impressed with the overall improvements at the station that the terrazzo floors of other stations in the area have now been similarly treated.

Source: HSE

Flooring at Hammersmith station (District and Piccadilly lines)

At Hammersmith, west London, a largely new London Underground (LUL) station was built in the mid-1990s. In the booking hall area the new flooring chosen was a ceramic tile. Unfortunately, the tiles failed LUL's acceptance testing on completion. An SRV of 40 in the wet was the requirement specified (the tiles had achieved this in factory testing).

After much debate, the agreed solution was to over-tile with similar tiles manufactured to a higher slip resistance. These passed the acceptance testing and remain in service.

The layout of the booking hall at Hammersmith is such that a large area is rarely used, with most passenger flow in direct lines between the ticket gates and the heads of stairs. Over the past 10 years the tiles have inevitably worn, as Hammersmith is a busy station, but the large areas out of the direct flow have retained their original high slip resistance.

Cleaning this booking hall is a real problem for the premises managers – the unused areas with the original rough surface profile retain the inevitable dust and grime of daily London life. Cleaning machinery gets damaged on this floor. The worn areas have lost this profile, but as this has worn down they have become easier to keep clean, contributing to an uneven appearance. The whole area scores badly against ambience measures, a disappointing performance for a "new" station.

Contrast between worn and less-trafficked floor tiles in Hammersmith station

Source: LUL

Airport carries out a wide-ranging review of pedestrian slip risks

Note: this is a detailed case study illustrating application of the principles of the SPM

A busy regional airport has enjoyed considerable growth in traffic over recent years. This growth has led to higher pedestrian traffic and on-site facilities have been developed to handle the increased passenger throughput. Airport management decided that it was an appropriate time to carry out a review of pedestrian slip risks on site. The review looked at risks to passengers, airport staff and those working for other organisations, such as the airlines and franchise-holders.

The review looked at a number of areas of the site where informed judgement suggested that there was at least the potential for slipping incidents. Micro-roughness measurements were taken of the floor surfaces in the areas under review. (Micro-roughness, in microns, is a significant factor influencing slip resistance of floors, especially in wet or contaminated conditions.) These measurements were considered together with other factors to provide a holistic picture and real-life assessment of slip risk.

Other slip risk factors incorporated into the overall assessment included:

- floor type and finish
- tasks and activities in the area under review
- who was using the area
- water or other likely floor contaminants
- staff footwear
- environmental influences in the area
- cleaning regimes.

A selection of the areas investigated, conditions found, conclusions drawn and responses made are summarised here.

Passenger check-in

At the passenger check pedestrians (including children and the elderly) are likely to be pushing, pulling or carrying loads, be rushing and possibly distracted. It has an epoxy-painted concrete floor that is regularly cleaned with a mechanical scrubber-drier. The entrance is well protected from the elements and has extensive built-in matting. Under normal conditions the check-in hall presented a low slip risk, but should water be walked in on wet days or accidental spillages inside not be attended to there would be a significant risk.

Response

The objective is a clean and dry walking surface. Monitoring to take place to see if walk-in water deposited; supplementary matting to be provided if necessary. Staff to be briefed on spillage-spotting and the action they should take.

Public concourse

Extending through a large part of the inside of the terminal, the floor has a textured epoxy surface. It is subject to the same sorts of conditions and use as the check-in. The matting is not so good in the concourse entrances, with some being ineffective because it is not placed to coincide with the direction that most pedestrians walk in after entering the door, so walk-in water is deposited as a result. Weather canopies are provided at most entrances to limit water contamination. The floor surface roughness is moderate and will not cope with dirty or heavily contaminated conditions.

Response

Routine cleaning systems to be reviewed to control contamination potential and to ensure that the floor is fully dry immediately following cleaning.

Continued overleaf

Airport carries out a wide-ranging review of pedestrian slip risks (*contd*)

Main revolving-door entrance

The principal passenger entrance to the terminal building is a large powered revolving door. The outer half of the floor within the revolving door is covered with ceramic tiles. These tiles have very low micro-roughness, are exposed to rain and dirt, and are located where encumbered pedestrians (young and old) are rushing, turning and distracted. The slip risk was identified as high. The inner half of the floor within the circumference of the revolving door consists of coarse grit/dirt-removal matting. Although this has some absorbent qualities, its design (and heavily worn condition) render it of little use in preventing walk-in water being carried on to the adjacent concourse, where control of contaminants is known to be important.

Response

Arrangements are being made to have the revolving door removed and pedestrian flow redirected to a better-controlled area. The decision was made easier by the "temperamental" nature of these particular powered doors.

Duty-free shop

The floor here is high-gloss ceramic decorative tiling with low micro-roughness. Originally thought to be likely to be a high-risk area, this proved not to be the case. The duty-free shop is sited in the heart of the terminal building, so there is no risk of walk-in water. Floor cleaning takes place at night when the shop is closed to the public. The policy forbidding food or drink being brought in to the shop is enforced. There are no open products to contaminate the floor and bottle breakage responses and procedures are very well organised.

Response

Although the floor is glossy, pedestrians never have anything other than a clean and dry floor to walk on. No action is needed.

Passenger toilets

Toilet facilities exist to serve the concourse in the land-side concourse area and the air-side departure lounge. All are cleaned to a high standard with a fully dry floor before being reopened to the public. However, all toilet areas are likely to be subject to water contamination of the floor and will be used by people of varying physical capability. The air-side toilet facilities have textured epoxy resin flooring offering high micro-roughness that does not impair cleaning but does offer high slip resistance. The land-side facilities have gloss terrazzo floors – a beautiful finish visually, but with very low micro-roughness and very slippery when wet.

Response

Air-side – slip risk is low even in foreseeable wet circumstances. No action. Land-side – a real likelihood of slips occurring. There would be difficulties in replacing the floor in advance of the next planned refit. Greater (labour-costly) frequency and diligence of splash clean-ups implemented. Etching treatment to the gloss terrazzo to improve surface micro-roughness and slip resistance in wet conditions being investigated.

Continued opposite

Airport carries out a wide-ranging review of pedestrian slip risks (*contd*)

Means of access to aircraft

Pedestrians boarding and alighting from aircraft can be expected to be carrying items, rushing and distracted. Some will be the young or elderly. Two main means of access exist: wheeled movable aircraft steps and powered "umbilical" air-bridges.

Movable steps

 These are kept outside and are exposed to the weather and a variety of potential contaminants to the walking surface, but the profiled rubber treads to the steps still offer reasonable slip resistance.

Response

Periodic cleaning of the treads to control contamination and monitoring of the condition of the rubber surface for hardening or perishing brought on by exposure to the elements.

Air-bridges

The air-bridges are floored with textured vinyl sheet for the most part and are cleaned to a high standard. Even with young and old, encumbered, rushing and distracted pedestrians, the textured vinyl – protected as it is from foreseeable contamination of any significance – performs well. The last few feet of the air-bridges (where they dock with the aircraft and where the operative stands) are floored with metal chequerplate. Experience and experimentation suggest that the slip resistance of metal chequerplate is not a direct function of the raised profiling pattern but related more closely to the micro-roughness of the upper surface (the foot contact surface) of each profile cleat. Micro-roughness measurements and analysis of the other relevant conditions indicate there is little problem if the chequerplate is kept clean and dry, but it offers poor slip resistance if wet or contaminated. The most likely source of contamination is rainwater if the docking roller doors of the air-bridge are opened early in the docking process. This is of particular concern as, if the roller door is opened early, the operative is standing on a moving slippery surface close to an open drop to the tarmac below.

Response

The high standards of floor cleaning to be maintained. Strict operating disciplines must be followed to prevent the air-bridge docking end door being opened until the last seconds of the docking manoeuvre.

Baggage handling hall

Where staff load luggage to and from trailers and conveyors in a handling bay opening on to the airport apron, the matt concrete surface has become ingrained with oily contamination and gets wet from pedestrians and baggage trucks from outside. Workers are manhandling loads and, although footwear standards are controlled, there is significant slip risk in this area. No real floor cleaning takes place.

Response

The original roughness of the concrete floor surface to be restored with a one-off rigorous clean. Routine mop and detergent cleaning to be introduced, with occasional use of a mechanical scrubber to prevent contaminants becoming ingrained and compromising the surface roughness and associated slip resistance.

Continued overleaf

Airport carries out a wide-ranging review of pedestrian slip risks (*contd*)

Airport apron

Used as access and thoroughfares by aircrew, passengers moving to and from smaller aircraft and by airport staff. The concrete pathways are sufficiently textured that, even in wet and contaminated conditions and with additional problematic human and environmental factors, they provide excellent slip resistance. However, walkways and other important points on the apron incorporate coloured lines marked out in a purpose-made paint-like material. These lines exhibit very low micro-roughness and involve a high slip risk in the unavoidably wet and contaminated conditions. (Motorcyclists will be familiar with the loss of grip when crossing broken white lines to overtake.)

Response

The apron markings have to be replaced or refreshed at intervals, so in the next renewal cycle alternative, possibly textured, surface marking products will be considered.

Airport fire station

The dock bays housing the airport's fire appliance are smartly kept. The floors of the appliance bays comprise light-coloured paint on concrete and are visibly clean and smooth – too smooth, in fact, as they have very low surface micro-roughness. Although fire crews' footwear can be controlled to a high standard, a regularly wet environment with shiny painted floors, cleaned by hosing down, detergents from appliance cleaning and inevitably hurried, distracted users indicate a significant slip risk. Getting injured on the way to the appliance is not a good way to respond to an emergency.

Response

The uses of and activities in the fire station dock bay seem fixed and largely unavoidable, but the floor paint is just starting to show its age in parts. Attempts to be made to identify a floor paint or other surface treatment with the required combination of cleanliness, durability and micro-roughness.

Mobile work equipment and workplace vehicles

Metal chequerplate is in use on many items of workplace transport and mobile work equipment at the airport, including fire appliances and working-at-height platforms (see also "Means of access to aircraft – air-bridges", above). It is generally used where people stand on the equipment to work or on footholds placed where staff climb up on to vehicles and equipment. The experiences of staff using these vehicles and equipment (usually outdoors and subject to water and other contamination) accord with the assessment that metal chequerplate can be particularly slippery when wet. Ad-hoc attempts have been made to tackle this known slip risk by applying adhesive anti-slip strips or sheets, although this has been far from comprehensive.

Response

Each vehicle and piece of equipment to be inspected and assessed for retrofitting of adhesive anti-slip coatings. "Fitting as standard" of chequerplate is to be raised as an issue when specifying for or procuring new vehicles or equipment.

Source: HSE, <www.hse.gov.uk/slips/experience/airport.htm>, accessed 1 Dec 2005

Oily floors in engineering workshop

For years many of the machines at a specialist engineering company leaked, which meant oil and sawdust were deposited continuously on the shopfloor.

Accidents to people slipping on the floor were reported at the rate of about 14 every year. For the size of the operation this represented quite a high incident rate.

What was done?

- Oil leak plan instituted – a list of leaking machines was prepared for maintenance staff to fix. This list is updated each time a leak is fixed or another leak is reported.
- Bunding – all machines with leaks that cannot be fixed (often older machines) have a metal bunding around the whole of the machine, which collects all the leaking oil to stop it spreading across the floor.
- Planned maintenance regime – each machine is serviced regularly, and part of the service requirement is to fix leaks.

Benefits of the oil leak plan

- Slip accident rates have been reduced by about two-thirds.
- The oil that is collected in the bunding is vacuumed out and recycled for reuse.
- Oil usage has been significantly reduced where leaks have been fixed, representing a big cost saving; additionally, less oil needs to be stored on site.
- Use of sawdust to soak up oil has been reduced significantly.
- Housekeeping has been greatly improved, resulting in a better working environment, improved motivation and greater incentive for staff to keep their work area tidy.
- A weekly housekeeping audit challenges supervisors of areas if they have machines with oil leaking on to the floor.
- Environmental impact – the company has ISO 14001 accreditation; audit results from the external auditor show a significant improvement in oil usage and reduction in oil leaks and risks of contamination.
- There is an action and improvement scheme whereby employees can pick a problem or suggested improvement and work on it as a team and implement it. If there are cost savings to the company then the team receive 10 per cent of the savings. Two of these worker initiatives have contributed to improving the handling of oil leaks, which has led to further improvements.

Source: HSE

Tackling slip problem in a manufacturing bakery

An HSE inspector received reports of workers being injured as a result of slip accidents at a factory producing filled pastry products. The inspector found that staff in manufacturing and ancillary areas were experiencing slip problems and that not enough was being done to prevent them. The inspector served an improvement notice on the company, requiring it to take remedial action. Although this legal notice was initial spur to action for the company, it went beyond the specific requirements of the inspector's notice to try to reduce the slip accident rate to zero.

Problems faced by the company included a variety of floor finishes on different parts of the site, the floors were often exposed to harsh wear and treatment, complete elimination of all floor contamination risks (water and ingredients) was not practical, and workers were often pushing or pulling loads. The company took advice from its HSE inspector and decided to introduce several changes to try to reduce slip risks.

Training was given to managers and the workforce in slip prevention, the co-operation of workers was also sought in assessing the effect of these changes.

Staff were issued with different types of "anti-slip" footwear. Some were initially assessed as performing well but were said to be difficult to keep clean as food debris clung to the tread pattern. Trials of one type of footwear were carried out using some of the male workers, but when these were subsequently issued to female workers they found them heavy on the feet and tiring. Another footwear type, which was initially felt not to have quite as high a slip resistance, seemed to deliver better results over time, as the tread pattern did not clog with food debris (better for hygiene) and offered greater comfort to many staff.

The factory had been in existence for many years and had changed hands. Not much was known about when floor finishes had been installed and why certain choices of floor finish had been made. An expensive floor repainting exercise had been carried out, but the finish was not durable. Epoxy "anti-slip" floor finishes had been installed in areas where there was perceived to be the greatest slip risk.

It transpired that the section felt to be most at risk because of the amount of water spillage – the tray-wash area – was not such a problem, because the micro-roughness of the floor surface provided good slip resistance without compromising hygiene. A greater risk was posed by the pastry preparation areas, where the floor surface roughness was insufficient to cope with the type and extent of contamination that was occurring.

Rather than looking at replacing the floor the company introduced measures (such as drip trays around machines and conveyors) to reduce and capture potential spillages – greatly reducing floor contamination – and introducing "anti-slip" footwear that did not clog. Important aspects of the slip prevention programme were the improved cleaning regimes designed to deal effectively with the food spillages that did reach the floor.

The company found that the number of slip accidents dropped to less than a quarter of those that they had been experiencing before the preventive measures were introduced.

Source: HSE, <www.hse.gov.uk/slips/experience/bakery.htm>, accessed 1 Dec 2005

Walkway level changes hard to spot

Alterations were made to walkways in a government office after an alert administration manager identified a risk to pedestrians.

The walkways were covered in dark maroon tiles and were in good condition throughout, but close to two entrance doors there were sloping areas of walkway and another area on a staff corridor. The building was quite old and the changes in level were part of the building structure, but, because of the dark colour of the floor, it was not obvious that there was a change in level. All of the areas were on fire-escape routes and one was at the entrance to the main reception. The manager knew that visitors could not be expected to know about the change in level and some visitors were likely to be less steady on their feet than others.

It would have been difficult to change the structure, and to remove the slopes by introducing a single step into the walkway would probably not have lessened the risk. The manager decided to highlight the slopes with contrasting, light-coloured areas of carpet tiles so that the changes in level attracted the attention and were apparent to pedestrians.

Although there had been no reports of accidents (neither injuries nor "near-misses") on these walkways, he decided that the very low cost of retiling these areas was worthwhile when compared with the potential cost of just one fall injury.

Sloped walkway not easy to spot at first glance (left) and with the change in level highlighted (right).

Source: HSE, <hse.gov.uk/slips/experience/walkway.htm>, accessed 1 Dec 2005

Refurbishment of university building

A recently refurbished university building provides an example of design that does not accord with the recommendations in this guide, giving rise to a number of deficiencies.

The campus is in a location subject to persistent wet weather. The entrance, always a vulnerable area in such buildings, is subject to high traffic load. It has only a small matting area and externally there is no protective canopy or covered access walkway. The photograph below indicates that there is also a high level of glare.

The tile selected by the designer was a sensible choice, but this was overruled by a senior member of the faculty who insisted on a less suitable product, which was in turn not backed up by appropriate manufacturer's literature. No specific cleaning regime was considered.

Measurements of slip resistance and surface roughness have been obtained, demonstrating that the slip resistance when the floor is wet is substantially reduced, giving rise to a significant slip risk.

Condition	Contamination	Test direction	SRV
As-found	Dry	Direction I	59
As-found	Dry	Direction II	45
As-found	Dry	Direction III	38
As-found	Dry	Direction I	44 (2nd tile)
As-found	Water wet	Direction I	14
As-found	Water wet	Direction II	14

Rz surface roughness: 3.8 micron (mean) (20 micron minimum recommended)

Pendulum SRV (Four-S rubber slider)

The internal ramp (pictured below) uses the same tiles as the entrance and is similarly subject to glare through the external glazing.

This example illustrates the danger of not following the recommendations given here for the selection and maintenance of walking surfaces. There is an obvious risk that the situation at the main entrance, if allowed to continue without even the benefit of improved matting or an appropriate management strategy, will lead to slip accidents occurring. These in turn have serious repercussions in terms of physical injury and financial costs to the university.

Source: HSE

HSL INVESTIGATIONS

The case studies below are taken from HSL reports produced in locations where slips had occurred and an investigation requested. The reports describe the assessment of floor slipperiness carried out and recommend action to be taken. For reasons of confidentiality, some details are omitted to avoid identification of people or buildings.

The reports emphasise that the results given relate to the as-found condition at the time of testing. They note that "factors such as changes in the nature or level of contamination, inappropriate cleaning/maintenance and long-term wear are known to have a profound effect on the slip resistance of floor materials".

12.3.1 Assessment of floor slipperiness in a hospital

Measurements of the pedestrian slip resistance of flooring were undertaken by a member of the HSL Pedestrian Safety Section during a joint HSL/HSE visit to the hospital in April 2003.

Three areas of flooring were assessed in and around the care of the elderly ward.

1. Bedroom 8, where a number of non-fatal injuries to patients had occurred in the recent past.

2. The en-suite bathroom to Bedroom 8.

3. A private bedroom sealed with a specialist anti-slip sealant.

Measurements in each case were carried out using standard HSL procedures.

Table 12.1 *HSL floor slipperiness assessment: elderly ward of a hospital*

Location	Area 1 (Bedroom 8)	Area 2 (en-suite shower room to Bedroom 8)	Area 3 (private bedroom)
Area	Central bedroom area	Adjacent to link doorway to bedroom.	At foot of bed.
Surface	Resilient flooring. Previously sealed with specialist product but sealant since stripped	Resilient safety (anti-slip) flooring, carborundum-based, no profiling.	Resilient flooring, sealed with specialist product. Highly reflective finish.
Typical use	Patients (elderly and/or with reduced cognitive function and/or incontinence); staff. Food and drink consumed in the area. Adjacent to en-suite bathroom.	Patients (elderly and/or with reduced cognitive function and/or incontinence); staff. Intentionally wet area, with en-suite toilet, wash-basin and non-guarded shower. Very high likelihood of regular wet contamination.	Patients (elderly and/or with reduced cognitive function); staff. Adjacent to en-suite bathroom.
Condition	All surfaces and seams (welded) in good condition. Some scratching evident. No contamination observed (localised or otherwise) but evidence of previous localised wet contamination.	All surfaces and seams in good condition. Low levels of localised contamination observed.	All surfaces and welded seams in good repair; some scratching evident. No contamination observed (localised or otherwise).
Testing	The surface roughness of the floor was measured. Testing of the CoF of the floor was carried out in the as-found (dry) condition, after the application of potable water using a hand-spray and after contamination with fine talcum powder.	The surface roughness of the floor was measured. Given the likelihood of localised contamination as noted above it was decided to test the CoF of the floor both before and after a simple clean-rinse-dry process. Further CoF testing of the lightly profiled surface in the "sit-in" shower cubicle was also undertaken.	The surface roughness of the floor was measured. Testing of the CoF of the floor was carried out in the as-found (dry) and in wet conditions.

Location	Area 1 (Bedroom 8)	Area 2 (en-suite shower room to Bedroom 8)	Area 3 (private bedroom)
Results	These indicated that the area under investigation presented an acceptably low slip risk in the clean, dry condition. The presence of low levels of wet contamination (eg beverage spills, cleaning residue, drips from the adjacent bathroom, unintentional biological contamination due to sickness, for example) resulted in unacceptable slipperiness. The presence of low levels of wet contamination on the soles of feet or footwear following use of the en-suite area would result in unacceptable slipperiness.	These indicated that the area under investigation presented a low slip risk in the clean, dry condition. The presence of low levels of wet contamination (water and water-based detergent solutions) resulted in unacceptable slipperiness. Although results suggested that the level of surface cleanliness could be improved upon, it was not thought that increasing either the frequency or the effectiveness of cleaning would significantly reduce slip risk. In the shower cubicle there was a moderate slip risk under contamination with potable water and detergent solution, both of which would routinely be found in use.	These indicated that the area under investigation presented an acceptably low slip risk in the clean, dry condition. The presence of low levels of wet contamination resulted in unacceptable slipperiness. The results obtained suggested that increasing the frequency or effectiveness of the cleaning regime would not reduce the slip risk appreciably.
Recommendations	If the potential of floor surface wet contamination is not reduced to negligibly low levels, modification or replacement of the floor surface should be undertaken.	The surface would be expected to be wet and it was not reasonable to expect that wet contamination could be controlled to reduce the slip risk. It was therefore recommended that the surface should be modified or replaced. In the shower area, it was suggested that in view of the regular use by those whose mobility may be impaired the slip resistance might sensibly be increased	It was felt that the flooring investigated was not suitable for use in an area where wet contamination might be experienced. The alternatives recommended were either the reduction of wet contamination to consistently near-zero levels, or modification or replacement of the flooring.

HSE guidance recommends that where action is needed to reduce the slipperiness of a given area, the hierarchy of control measures is, in simplified form, as follows:

- the reduction of floor surface contamination to negligible levels
- substitution of high-viscosity contaminants with those of lower viscosity
- the adoption and issue of specialist anti-slip footwear
- floor surface modification, or
- floor surface replacement (HSE, 1996d).

It is obviously necessary to consider the viability of these measures in relation to the particular usage and circumstances of the area under review. The demography of the patients using the care of the elderly ward, and the related risk of accidental spills and biological contamination, suggests that a strategy relying on contamination control would not possible in Areas 1 or 2, although this would be appropriate in Area 3.

It is equally impracticable to control the footwear used in and around Bedroom 8 and the en-suite area, meaning that either modification of the flooring or its replacement should be considered. For the particular materials concerned, simple modification is not effective and replacement is therefore the recommendation, using an appropriate product with a suitable anti-slip layer.

This case study demonstrates how, in cases where testing indicates that a reduction in slipperiness is required, the application of control measures needs to take account of the particular circumstances of use. This is also an example where the flooring is found to be inappropriate for its use.

12.3.2 Assessment of floor slipperiness in a supermarket

Measurements of the pedestrian slip resistance of flooring were undertaken by members of the HSL during a site visit undertaken at the request of an environmental health officer from the local district council. The HSL report was issued in May 2004.

The installed flooring was tested at the site of a serious slip incident. Slip resistance assessments were carried out in accordance with the guidelines recommended by the United Kingdom Slip Resistance Group (UKSRG, 2000). Two test methods were used, a pendulum coefficient of dynamic friction (CoF) test and an Rz surtronic micro-roughness transducer.

Table 12.2 *HSL floor slipperiness assessment: supermarket*

Area	Shop floor – accident site
Surface	Smooth vinyl tile
Typical use	Regular staff and general public pedestrian use
Condition	Some apparent wear, deep scratching present, clean appearance
Testing	Pendulum CoF data were collected from the site of the reported accident: ● in the as-found, dry state ● in a contaminated state after the application of potable water by hand-spray ● after cleaning implemented using the in-house regime. Surface micro-roughness measurements were taken on seven separate tiles, at and adjacent to the accident site and in adjacent aisles.
Results	In the as-found dry condition the slip resistance values suggest an extremely low potential for slip. Application of a fine mist spray results in a high slip potential, while the cleaning regime results in a slight increase in slipperiness. Inadequate rinsing, ie the retention of detergent, may be partly responsible for the high slip potential.
Recommendations	In wet or contaminated conditions the existing flooring presents an unacceptably high slip risk. The risk is unacceptably high even with small volumes of clean water. It is therefore considered that use of the primary control measures in the HSE guidance would be unlikely to reduce the slip risk to an unacceptably low level. Given that there could be only partial control over the use of footwear in the area it is recommended that the slip resistance of the flooring material in the area investigated should be improved.

The recommendation in this case takes account of the sharp increase in slip risk resulting from even low levels of wet contamination. It also recognises that in a commercial area where the public has primary use it is not practicable to exercise control over footwear as a means of reducing slip risk (or indeed for other purposes).

Investigation of floor slipperiness in a fast-food restaurant kitchen

A full forensic slipperiness investigation of a fast-food restaurant kitchen following a slip incident was undertaken during a joint HSL/local authority visit in July 2004. The slip resulted in a serious burn, because a deep-fat frier was pulled over on to the injured person. The investigation was carried out using standard HSL/HSE techniques in accordance with the guidelines issued by the UKSRG where applicable.

Two test methods were used, a calibrated Stanley pendulum slipperiness assessment instrument and a calibrated Surtronic Duo surface micro-roughness transducer set to the Rz parameter.

Table 12.3   *HSL floor slipperiness assessment: fast-food restuarant kitchen*

Floor type	Tile type A	Tile type B	Tile type C
Surface	Pressed ceramic replacement tile with light carborundum effect	Pressed ceramic replacement tile with light carborundum effect	Pressed ceramic tile (original) with light carborundum effect
Condition	Worn very smooth	Worn smooth	No comment
Testing	Pendulum tests were undertaken: • in the as-found dry condition • after application of low volumes of potable water by hand-spray • on an oil-contaminated surface • after using an existing cleaning technique with mop and bucket, surfaces mop-wet. Two test slider materials were used. Surface roughness measurements were also undertaken for each tile type.		
Results	The SRV values suggest that this surface is moderate-to-high risk in as-found dry conditions, and high risk when wet or contaminated.	The SRV values suggest that this surface is moderate-to-high risk in as-found dry conditions, and high risk when wet or contaminated	The SRV values suggest that this surface is moderate-to-high risk in as-found dry conditions, and high risk when wet or contaminated
Mean Rz surface roughness	8.8 μm	9.8 μm	13.6 μm
Recommendations	• Consider footwear trials • consider use of matting in some areas, but note this could provide a trip hazard • if neither of these strategies is successful then alternatives are, initially, modification of the surface by acid-etching or replacement of the flooring.	• Consider footwear trials • consider use of matting in some areas, but note this could provide a trip hazard • if neither of these strategies is successful then alternatives are, initially, modification of the surface by acid-etching or replacement of the flooring.	• Effective cleaning of the floor with subsequent drying may be adequate • consider footwear trials • consider use of matting in some areas but note this could provide a trip hazard • if none of the above strategies is successful then alternatives are, initially, modification of the surface by acid-etching or replacement of the flooring.

Note that the surface roughness values for the two replacement tiles are similar. Both lie within the "high" category for potential for slip (<10 for water-wet, low-pedestrian-activity areas). The original tile, in contrast, lies within the "moderate" range of between 10 and 20.

This suggests that the replacement tiles might be considered to be of an inappropriate specification for the intended activity use.

Investigation of floor surface slipperiness at a leisure centre

A full forensic slipperiness investigation of a leisure centre following a reported serious accident was undertaken during a joint HSL/HSE visit in October 2004. The reported incident involved an infant slipping. It was unclear, however, exactly where the slip occurred. Accordingly investigations were undertaken on the four surfaces in the area implicated, together with supplementary measurements on the tiles in the main changing room and at the poolside. The investigation was carried out using standard HSL/HSE techniques in accordance with Issue 2 of the UKSRG guidelines (UKSRG, 2000) where applicable.

Two test methods were used, a calibrated Stanley pendulum slipperiness assessment instrument and a calibrated Surtronic Duo surface micro-roughness transducer set to the Rz parameter. Required slip-resistance values were calculated for slopes of 2.5° and 5° to allow interpretation of the pendulum data generated on site.

Table 12.4   *HSL floor slipperiness assessment: leisure centre*

Location		A	B	C	D	E	F
Surface		Stainless steel drain cover on ramp	Pressed ceramic profile tile on slope	Pressed ceramic profile tile on slope	Pressed ceramic profile tile on slope	Pressed ceramic profile tile in changing room	Poolside tiles – "cratered appearance"
Incline		4.2°	2.7–3.7°	3.6°	0.6°	n/a	n/a
Mean Rz surface roughness		1.9 µm	29.0 µm low traffic 24.7 µm high traffic	56.6 µm low traffic 60.5 µm high traffic	42.2 µm low traffic 33.2 µm high traffic	80.5 µm	27.3 µm
Pendulum testing		Pendulum tests were undertaken: • in the as-found, dry condition • after application of low volumes of potable water to the flooring by hand-spray Two test slider materials were used, Four-S rubber and TRRL rubber Data was generated in three complementary test directions in order to account for the presence of floor surface directionality (a variety of profiled tiles was used).					
Results	Pendulum data using the Four-S rubber slider	Unacceptably **high** slip risk in water-wet conditions	**Low** slip risk to shod pedestrian traffic in water-wet conditions	**Moderate** slip risk to shod pedestrian traffic	**Low** slip risk to shod pedestrian traffic	**Moderate** slip risk to shod pedestrian traffic	**Moderate** slip risk to shod pedestrian traffic
	Data from the micro-roughness meter			**Low** slip risk in water-wet conditions	**Low** slip risk in water-wet conditions	**Low** slip risk in water-wet conditions	**Low** slip risk in water-wet conditions
	Data from the TRRL slider	–	Unacceptably **high** slip risk in water wet conditions to barefoot or trainer-shod pedestrians	Unacceptably **high** slip risk to barefoot or trainer-shod pedestrians	**Moderate** slip risk to barefoot or trainer-shod pedestrians	**Moderate** slip risk to barefoot or trainer-shod pedestrians	**Moderate** slip risk to barefoot or trainer-shod pedestrians

Discussion

To address the problems in existing premises, it is recommended that the following control measures be considered (in no particular order):

- prevention of contamination
- management of spillages and cleaning regimes
- effective matting systems
- choice of suitable footwear
- design of workplace and work activities maintenance of plant and the work environment
- specification of appropriate flooring
- housekeeping
- effective training and supervision
- clearly mark the beginning and end of slopes using colour contrast.

The list of factors affecting the slip risk should be used to form an overall management approach to the pool and changing room areas. The factors outlined will interact with one another and are not mutually exclusive. It is considered in this case that cleaning and floor specification are the most significant factors, as there is little control over the traffic or the levels of contamination. It is suggested that cleaning trials be undertaken and their effectiveness measured using the techniques outlined above. If cleaning is unsuccessful, floor surface modification or replacement may be necessary.

Appendices

A1   Standards on testing

A1.1   DIN RAMP TESTING

DIN 51097:1980. *Testing of ceramic floor coverings. Determination of the anti-slip property. Wet use areas walked on by bare feet*

DIN 51130:1992. *Testing of floor coverings. Determination of the anti-slip properties. Workrooms and fields of activities with raised slip danger. Walking method – ramp test*

A1.2   PENDULUM TESTING

BS 812-114:1989. *Testing aggregates – method for determination of the polished-stone value* (superseded)

BS EN 1097-8:2000. *Tests for mechanical and physical properties of aggregates. Part 8: Determination of polished stone value*

BS 6717:2001. *Precast, unreinforced concrete paving blocks – requirements and test methods*

BS 7263-1:2001. *Precast concrete flags, kerbs, channels, edgings and quadrants. Part 1: Precast, unreinforced concrete paving flags and complementary fittings – requirements and test methods*

BS 7923:2003. *Determination of the unpolished and polished pendulum test value of surfacing units*

BS 7976-1:2002. *Pendulum testers. Part 1: Specification*

BS 7976-2:2002. *Pendulum testers. Part 2: Method of operation*

BS 7976-3:2002. *Pendulum testers. Part 3: Method of calibration*

BS EN 1338:2003. *Concrete paving blocks – requirements and test methods*

BS EN 1339:2003. *Concrete paving flags – requirements and test methods*

BS EN 13036-4:2003. *Road and airfield surface characteristics – test methods. Part 4: Method for measurement of slip/skid resistance of a surface – the pendulum test*

BS EN 1341:2001. *Slabs of natural stone for external paving – requirements and test methods*

BS EN 1342:2001. *Setts of natural stone for external paving – requirements and test methods*

BS EN 1344:2002. *Clay pavers – requirements and test methods*

DD ENV 12633:2003. *Method of determination of unpolished and polished slip/skid resistance value*

A1.3   SURFACE ROUGHNESS

BS 1134-1:1988. *Assessment of surface texture. Part 1: Methods and instrumentation*

A1.4   SLED-TYPE TESTING

FSC2000. *Sled test*

GMG100.

A1.5   TESTING FOR ROAD SURFACES

BS EN 1097-8:2000. *Tests for mechanical and physical properties of aggregates. Part 8: Determination of the polished stone value* (supersedes BS 812-114:1989)

A1.6   AMERICAN TESTING

A1264.2-2001. *Standard for the provision of slip resistance on walking/working surfaces* (ANSI/ASSE)

ASTM C1028. 1996. *Test method for determining the static coefficient of friction of ceramic tile and other like surfaces by the horizontal dynamometer pull-meter method*

ASTM F 1678. 1996. *Test method for using a portable articulated strut slip tester (PAST)*

ASTM F 1679. 2004. *Test method for using a variable incidence tribometer (VIT)*

Americans with Disabilities Act accessibility guidelines (1990)

NFSI 101-A (National Floor Safety Institute)

A1.7   AUSTRALIAN AND NEW ZEALAND TESTING

AS 4586. 2004. *Slip resistance classification of new pedestrian surface materials*

AS 4663. 2004. *Slip resistance measurement of existing pedestrian surfaces*

Standards Australia handbook HB 197. *An introductory guide to the slip resistance of pedestrian surface materials* (1999)

A2   Legislation

A large body of regulation applies to health and safety within the UK. This is contained within the Building Regulations and occupational health and safety regulations. There are separate systems of Building Regulations within England and Wales, Scotland and Northern Ireland. Similar intent exists under all three systems, but with variations in both the scope and method of application. This guide cites references to the system in England and Wales. Building Regulations within Northern Ireland and Scotland can be found online at, respectively: <www.dfpni.gov.uk/buildingregulations/technical.htm> and <www.sbsa.gov.uk>. Some of the principal instruments of particular relevance to controlling the risk of slips are identified below, with a brief explanation of their scope.

A2.1   BUILDING REGULATIONS 2000

Approved Document M *Access to and use of buildings*

Access and use

M1 Reasonable provision shall be made for people to:

- gain access to; and
- use

the building and its facilities.

Access to extensions to buildings other than dwellings

M2 Suitable independent access shall be provided to the extension where reasonably practicable.

   1.6 As far as possible, access should be level from the boundary of the site, and from any car parking designated for disabled people, to the principal entrance and any entrance used exclusively for staff...

   1.13 A "level approach"... will satisfy Requirement M1 or M2 if:

      d. its surface is firm, durable and slip resistant, with undulations not exceeding 3 mm under a 1 m straight edge for formless materials.

      e. where there are different materials along the access route they have similar frictional characteristics.

A ramped access will satisfy Requirement M1 or M2 if:

- the ramp surface is slip-resistant, especially when wet, and of a colour that contrasts visually with that of the landings
- the frictional characteristics of the ramp and landing surfaces are similar

Materials for treads should not present a slip hazard, especially when the surface is wet.

Scotland

Domestic and non-domestic technical handbooks provide guidance on achieving the standards set out in the Building (Scotland) Regulations:

- Section 2 *Fire* includes means of escape and emergency lighting
- Section 4 *Safety* includes access to and within buildings and stairs and ramps.

HEALTH AND SAFETY AT WORK ETC ACT 1974

This is the umbrella health and safety legislation in the UK. It imposes broad duties on employers, employees, the self-employed and those in charge of premises. A suite of regulations covering particular aspects of health and safety in more detail comes under this Act. Some of the regulations have been enacted to bring UK health and safety law into line with European directives. The CDM Regulations followed an EC directive, and the Workplace (Health, Safety and Welfare) Regulations are another example.

The HSW Act requires employers to ensure the health and safety of their employees and others who may be affected by their work activities. This includes acting to control slip and trip risks. Under the HSW Act employees must not endanger themselves or others, must co-operate with their employer and use any safety equipment provided by their employer. Manufacturers and suppliers have a duty to ensure that their products are safe. Adequate information about the appropriate use of products must also be provided. More recent regulations emphasise the importance of such measures.

Section 2

The duty of the employer is to ensure, so far as is reasonably practicable, the health, safety and welfare at work of all his employees. In relation to the avoidance of slips, this duty is met primarily by:

(d) so far as is reasonably practicable as regards any place of work under the employer's control, the maintenance of it in a condition that is safe and without risks to health and the provision and maintenance of means of access to and egress from it that are safe and without such risks;

(e) the provision and maintenance of a working environment for his employees that is, so far as is reasonably practicable, safe, without risks to health, and adequate as regards facilities and arrangements for their welfare at work.

Section 7

It shall be the duty of every employee while at work to take reasonable care for the health and safety of himself and of other persons who may be affected by his acts or omissions at work; regarding requirements imposed upon the employer to co-operate with him so far as is necessary to enable that duty to be complied with.

MANAGEMENT OF HEALTH AND SAFETY AT WORK REGULATIONS 1999

The Regulations impose a duty on the employer to assess the risks to their employees and others and to identify and put into place measures to reduce the risks to the required level. They require employers to assess risks (including slip and trip risks) to their employees and others (contractors, visitors etc) arising from work activities. Employers should be able to demonstrate they have effectively considered the risks and instituted suitable control measures. They also need to ensure that the measures they have taken are effective. They should investigate any significant slip and trip incidents.

Employees have a duty to report any situation that might present a serious and imminent danger and they should also notify employers of any shortcomings in the health and safety arrangements.

The above will apply generally to duty-holders and employees.

Regulation 3 imposes a duty on employers and self-employed people to carry out assessments of the risks to the health and safety of their employees and any others who may be affected by their work or business. Such assessments are required for the purpose of identifying any measures needed for compliance with health and safety law. The risks referred to would include slip (and trip) risks, and the measures referred to would be required to mitigate such risks.

There is an accompanying approved code of practice (ACoP) to the MHSW Regulations and its associated guidance *Management of health and safety at work*, published by HSE.

A2.4   WORKPLACE (HEALTH, SAFETY AND WELFARE) REGULATIONS 1992

Condition of floors and traffic routes

12. (1) Every floor in a workplace and the surface of every traffic route in a workplace shall be of a construction such that the floor or surface of the traffic route is suitable for the purpose for which it is used.

    (2) Without prejudice to the generality of paragraph (1), the requirements in that paragraph shall include requirements that:

       (a) the floor, or surface of the traffic route, shall have no hole or slope, or be uneven or slippery so as, in each case, to expose any person to a risk to his health or safety; and

       (b) every such floor shall have effective means of drainage where necessary.

    (3) So far as is reasonably practicable, every floor in a workplace and the surface of every traffic route in a workplace shall be kept free from obstructions and from any article or substance which may cause a person to slip, trip or fall.

    (4) In considering whether for the purposes of paragraph (2)(a) a hole or slope exposes any person to a risk to his health or safety:

       (a) no account shall be taken of a hole where adequate measures have been taken to prevent a person falling; and

       account shall be taken of any handrail provided in connection with any slope.

Guidance on what action can be taken can be found in the approved code of practice (para 89 to 105). Essentially the floor should:

- be of sound construction, have adequate strength and stability for the loads placed upon the floor or passing over

- not be overloaded

- have the surfaces free from holes, slopes or uneven or slippery surface which is likely to cause a slip, trip or fall

- have a surface that will not cause a person to drop or lose control of anything being lifted or carried

- not cause instability or loss of control of vehicles or their loads

- be capable of containing spillages and leaks by bunding and draining away.

The Workplace (Health, Safety and Welfare) Regulations 1992 apply to *existing* workplace premises as well as new ones. **EC Directive 89/654/EEC** was transposed in UK law through the Workplace (Health, Safety and Welfare) Regulations 1992.

Commentary

The requirement for floors not to be slippery (12(2)) is a high standard as it is not qualified by the principle of "as far as reasonably practicable" (AFARP), and this poses the question of how slipperiness is assessed. 12(3) is not absolute, but refers to contamination.

Floor contamination

The W(HSW) Regulations require that floors should be kept clear of contamination AFARP. Often this can be achieved fairly easily and cheaply. For example, workers entering the stores from outside take a short cut through a workshop, trailing rainwater in bad weather and causing a slip hazard – a practice that could be cut out with a minimum of effort and goodwill.

Similarly, leaks of lubricant on to the floor from machinery may be fairly easy to prevent or contain, through good maintenance and fitting of a drip tray – much less effort than operating a system to deal with the spillages. Employers should also consider slips risks from solid contaminants, eg powders, granules, even plastic wrapping. It is important that employers have a system to clean up spillages straight away and have a cleaning regime that is appropriate to the type of floor and contaminant. Inadequate drying following cleaning can be hazardous. It may be necessary to consider the training of the cleaners (HSE, 2003a).

A2.5   PROVISION AND USE OF WORK EQUIPMENT REGULATIONS 1998

Application of requirements under these Regulations

4.   (1) The requirements imposed by these Regulations on an employer shall apply in respect of work equipment provided for use or used by any of his employees who is at work or who is on an offshore installation within the meaning assigned to that term by Section 1(4) of the Offshore Safety Act 1992.

   (2) The requirements imposed by these Regulations on an employer shall also apply:

   (a) to a self-employed person, in respect of work equipment he uses at work

   (b) to any person who has control, to any extent, of non-domestic premises made available to persons as a place of work, in respect of work equipment used in such premises by such persons and to the extent of his control; and

   (c) to any person to whom the provisions of the Factories Act 1961 apply by virtue of Section 175(5) of that Act as if he were the occupier of a factory, in respect of work equipment used in the premises deemed to be a factory by that section.

Any reference in paragraph (2)(b) to a person having control of any premises or matter means the person having control of the premises or matter in connection with the carrying on by him of a trade, business or other undertaking (whether for profit or not).

Suitability of work equipment

5.   (1) Every employer shall ensure that work equipment is so constructed or adapted as to be suitable for the purpose for which it is used or provided.

   (2) In selecting work equipment, every employer shall have regard to the working conditions and to the risks to the health and safety of persons which exist in the premises or undertaking in which that work equipment is to be used and any additional risk posed by the use of that work equipment.

   (3) Every employer shall ensure that work equipment is used only for operations for which, and under conditions for which, it is suitable.

   (4) In this regulation "suitable" means suitable in any respect which it is reasonably foreseeable will affect the health or safety of any person.

A2.6 CONSTRUCTION (DESIGN AND MANAGEMENT) REGULATIONS 1994 (CDM REGULATIONS)

These regulations apply to refurbishment work and to new-build. Designers are required under Regulation 3 to "prepare designs with adequate regard to health and safety" (ACoP paragraph 112). Paragraph 128 of the ACoP (HSE, 2001) notes that designers "need to understand how the structure can be constructed, cleaned and maintained safely" – although "cleaned" here does not include floors. The draft guidance *Managing health and safety in construction* (2005) in paragraph 133 also refers to designers' responsibilities as covering "the health and safety of users of workplaces".

A major refurbishment or new building work is an opportunity to eliminate slip and trip hazards. The work may be subject to the CDM Regulations.

A2.7 SAFETY REPRESENTATIVES AND SAFETY COMMITTEES REGULATIONS 1977

Safety representatives appointed under these regulations must be consulted on health and safety matters. They must also be given access to information relevant to the health and safety of the workers they represent, including any information relating to potentially hazardous conditions, such as slip and trip risks.

A2.8 HEALTH AND SAFETY (CONSULTATION WITH EMPLOYEES) REGULATIONS 1996

These require employers to consult with workers, either directly or indirectly through elected representatives, on matters relating to their health and safety at work. Safety representatives can help employers with both the development and implementation of a slip and trip risks policy. They will be able to identify risks in the workplace and bring the workers' perspective to the policy-making process.

A2.9 DISABILITY DISCRIMINATION ACT 1995

The DDA was passed in 1995 and parts of it became law for employers in December 1996. Since then other parts have been introduced, and since 1 October 2004, through Part 3, "providers of services" (ie businesses and organisations) have had to make reasonable adjustments to the physical features of their premises to overcome physical barriers to access, in particular Section 21 sub-section (2). In assessing the slip resistance of a floor, the use of the building (or other area) has to be considered.

Disability is defined as a physical or mental impairment that has a substantial and long-term adverse effect on a person's ability to carry out normal day-to-day activities. These activities include those relating to:

- mobility
- physical co-ordination
- ability to lift, carry or otherwise move everyday objects
- perception of the risk of physical danger.

A2.10   OCCUPIER'S LIABILITY ACTS 1957 AND 1984

These acts place common law duty-of-care responsibilities on owners in respect of visitors (1957) and others who are not visitors (trespassers) (1984). The duty of care to children is higher than that for adults. The extent of care will be determined by the circumstances at the time of any event. Warning signs would be included in the assessment of the adequacy of any preventative measures when determining if the owner had done all that was reasonable in the circumstances.

A2.11   CONTROL OF SUBSTANCES HAZARDOUS TO HEALTH REGULATIONS 2002 (COSHH)

While the sections above cover obligations in relation to reducing the risk of slip by the provision of appropriate surfaces, there are further regulations dealing with risks associated with their maintenance. Some of the cleaning agents used to ensure that surfaces are maintained at the required level of cleanliness and hygiene, and also with a low level of slip risk, may themselves be hazardous to the health of the cleaning operatives and building users. It is essential that they are used strictly in accordance with manufacturers' instructions and any local regulations, that appropriate protective clothing is worn and that they are both stored and disposed of also in accordance with manufacturers' instructions.

The COSHH Regulations cover the use of chemicals or other hazardous substances at work that can put people's health at risk. The law requires employers to control exposure to hazardous substances to prevent ill health. They have to protect both employees and others who may be exposed by complying with the Regulations.

COSHH is a useful tool of good management that sets eight basic measures that employers, and sometimes employees, must take. Effects from hazardous substances range from mild eye irritation to chronic lung disease or, on occasions, death.

Regulation 6 requires that an employer shall not carry on any work that is liable to expose employees to any substance hazardous to health unless he has made a suitable and sufficient assessment of the risks created by that work to the health of those employees and of the steps that need to be taken to meet the requirements of the regulations. Furthermore, regulation 7 requires that every employer shall ensure that the exposure of his employees to substances hazardous to health is either prevented or where this is not reasonably practicable, adequately controlled.

Flooring specifications may well include the use of materials for which COSHH data sheets are required from the manufacturer, eg free isocyanate monomer present in polyurethane resin-based floors and fine alkaline particles found in cement-based flooring products. The construction process itself may also require a risk assessment to be undertaken for possible hazardous materials, for example where the scabbling of an existing floor surface is required to remove the floor finish and contamination within the sub-base material to provide an adequate surface for the new flooring material.

A2.12   NATIONAL HEALTH SERVICE AND COMMUNITY CARE ACT 1990

Section 60 and Schedule 2 paragraph 18 provide that no health service body is a servant or agent of the crown and does not enjoy crown immunity except as expressly provided for by the Act. Therefore, except where the Act expressly provides, health authorities and NHS trusts are expected to comply with legislation such as the HSWA, and are as vulnerable to the consequences of failure to comply as other organisations (see NHS *Finance manual*). This is an important driver in the NHS organisation.

A3 Statistics

A3.1 GENERAL

The statistics for slips and trips (and often falls) (STFs) are generally quoted together. They are given in a variety of sources, but while the numbers may differ slightly in different industries and in different countries, the overall message is very consistent:

- they account for a very high proportion of all workplace accidents
- the cost is high – to the employer and the individual
- the long-term consequences are often serious (and may be fatal – see also Chapter 1).

STF accidents have accounted for more than 34 000 injuries each year since 1995 (HSC, 2001).

They are estimated to cost the UK economy as a whole more than £1 billion a year.

More than 42 per cent of all major accidents in the LA-enforced sector are STF on the level, and 23 per cent of over-three-day injuries (HELA, 2001).

Around 95 per cent of major injuries resulting from a slipping incident involve fractures and dislocations with sprains and strains accounting for over half the reported over-three-day injuries (*Contract Flooring Journal*, October 2002).

Slips and trips are the most common cause of major injuries at work. They occur in almost all workplaces, 95 per cent of major slips result in broken bones and they can also be the initial causes for a range of other accident types such as falls from height.

Slips and trips

Statistically – on average – they account for

- 33 per cent of all reported major injuries
- 20 per cent of over-three-day injuries to employees
- two fatalities per year (it is likely that the number attributable is much higher than the figure reported)
- 50 per cent of all reported accidents to members of the public
- cost to employers £512 million per year
- cost to health service £133 million per year
- incalculable human cost.

This means 1 slip or trip accident occurs every 3 minutes!

Source: HSE, <www.hse.gov.uk/slips/index.htm>, accessed 1 Dec 2005

A3.2   STATISTICS FROM OVERSEAS

Information from other countries shows a similar pattern of both incidents and concerns. The following was included in the conference "International network on the prevention of accidents and trauma at work", held by the European Agency for Safety and Health at Work at Helsingør, Denmark, 3–6 September 2002.

Trips, slips and falls in SMEs – analysing risks in the meat industry

Accidents at work involving trips, slips and falls (TSFs) happen very frequently and the resulting costs are high, both for employers and employees. In certain types of small and medium-sized firms (SMEs), such as those in the meat processing industry, trips, slips and falls account for more accident compensation payments than other accidents

TSF accidents in the German industry and trade

In the German industry and trade about 225 000 injuries in the workplace per year are caused by trips, slips, and falls and result in absences of over three days (>3 days) from the workplace. These kinds of accidents are usually known as "TSF accidents". TSF accidents are responsible for 19 per cent of all >3-days absences from the workplace and 25 per cent of all annually new accident compensation payments. The risk of getting seriously injured in a TSF accident is 1.4 times higher than for all other types of accidents.

TSF accidents in the German meat industry

The German meat industry reports that approximately 3000 TSF accidents per year make up 12 per cent of all absences of more than three days from the workplace. This is quite a low level for all of the industry, but, given that TSF accidents account for 50 per cent of all new accident compensation payments annually, it means that in the meat industry there is a higher risk of getting seriously injured from a TSF accident. In fact, you are four times more likely to be seriously injured from a TSF accident than from all other types of accidents

Causes of TSF accidents

An evaluation of the detailed accident questionnaire indicated that about 50 per cent of the TSF accidents were caused by technical/structural reasons and the remaining 50 per cent were due to the behaviour of the victims. The percentage of technical/structural reasons is relatively high. The result could have been influenced by the opinion of the victims. But an analysis of the accidents showed that more than 53 per cent of stairways had defects and 80 per cent were poorly lit. The study of the friction coefficients of the floor surfaces indicated particularly low values in the refrigeration rooms. Similarly, a study of the footwear worn during the accidents showed that these often remained in use long after their recommended life cycle.*

Relative duration of stay compared to frequency of TSF accidents for the sales staff

The sales staff spend 92 per cent of their working time in work areas with dry surfaces. If they leave this area, the risk rate can be more than three, and up to four, times higher in production and refrigeration rooms. The above analysis shows that the accident risk increases once the employee leaves their usual work area.

Structural and organisational measures are necessary to avoid staff moving between different workplaces with different flooring, particularly between dry and wet areas.

Source: Kloss (nd)

*   Schenk and Selge (1999). *Prävention von stolper-, rutsch- und sturzunfällen in betrieben der fleischwirtschaft*. Erich Schmidt Verlag, Bielefeld

According to the US Government, 265 000 workers in the United States sustained non-fatal injuries from slips, trips and falls in 2003, each resulting in one or more days away from work (BLR, 2005).

Avoiding slips, trips and falls in the office

Slips, trips, and falls constitute the majority of general industry accidents. In the US, they cause 15 per cent of all accidental, job-related deaths and are second only to motor vehicles as a cause of fatalities, according to the US Occupational Safety and Health Administration (OSHA).

Source: AT&T (2005)

A similar picture emerges from Australia.

> *Workers compensation statistics show that 16 per cent of all injuries result from falls on the level – this constitutes three injuries per 1000 workers per year. Falls of the same level – which may be caused by slips or trips – have been the cause of 11 per cent of back injuries. Back injuries sustained in this manner are likely to be more severe than those caused by over-exertion.*

> (WorkCover, 1998a)

The average cost of s slip, trip or fall accident was given as A$12 000. This figure is itself significant, while the hidden costs to the employer can be as much or more than the compensation figure. A related document, the Guide to preventing slips, trips and falls, produced by the Australian Government, notes the following.

> *Workers' compensation insurance covers only a small proportion of the costs associated with an injury or illness. It does not account for indirect costs such as:*
>
> - *the time to process and manage the injury*
> - *increased workloads for other staff to perform the injured person's work*
> - *loss of expertise and necessary skills*
> - *additional training needed for replacement staff*
> - *decreased productivity*
> - *the human aspect of pain and suffering.*

> (ComCare, 2002)

Statistical information is also produced on an industry and sector-related basis. The HSE statistics unit provides information showing percentages, rates and industry sectors for slips and falls.

FOOD AND CATERING

HSE Catering Information Sheet No 6 (HSE, 2005b)

Slips and trips risks are especially important to caterers because:

- injuries are frequent (about twice the average rate for manufacturing industry)
- they are the main cause of both reported serious injury (75 per cent) and all injuries (33 per cent)

HSE Food Information Sheet No 6 (HSE, 1996c)

Slips and trips risks are especially important in the food industry because:

- they occur four times more often than the average for industry, and are the main reason for the relatively high overall injury rate in the food industry
- they are the largest cause of serious injury (32 per cent) in the industry; there is a high rate in all sectors
- the potential losses could be significant, including costs (estimated at £22 million annually to employers in food, drink and tobacco); loss of key staff; liability (compensation, legal costs, insurance premiums and enforcement action); and individual suffering and disability.

Research from Australia reports that in the catering industry, which has a very high rate of slip-related injuries, 95 per cent of the slips occur on wet surfaces (Bowman, nd).

A3.4   EDUCATION

HSE Education Information Sheet No 2 (HSE, 2003b)

Introduction

Slips, trips and falls on the level are the most common cause of major injuries in workplaces and the second highest cause of over-three-day injuries. They occasionally cause fatalities, for example from head injuries.

The financial costs of slip and trip incidents are considerable. Based on 1999/2000 figures it is estimated that they cost employers £368 million and society as much as £763 million.

Slips and trips in education

Although slips and trips can happen to anyone, it is older people, particularly women, who are often injured more severely. A simple slip can even lead to death.

HSE statistics suggest that slips and trips are a major cause of accidents to education employees, pupils/students and others (see Table 1).

Slip and trip incidents can be controlled, provided sufficient attention is given to the nature of the work environment and the organisation affords them sufficient importance. The control measures needed are often simple and low-cost, but will bring about significant reductions both in human suffering and costs.

Table A3.1   *Slip and trip incidents in education for 1999/2000*

	Members of the public*			Employees		
	Reported injuries**	Injuries due to slips and trips	Percentage of injuries due to slips and trips	Reported injuries**	Injuries due to slips and trips	Percentage of injuries due to slips and trips
Primary and secondary education	6898	2296	33%	4032	1382	34%
Higher and further education	1484	358	24%	2093	599	29%

*   ie pupils, students and visitors
**   fatal and major injuries
***   fatal, major and over-three-day injuries

A3.5   HEALTH SERVICES

HSE Health Services Sheet No 2 (HSE, 2003c)

More than 2000 injuries to employees in healthcare, attributed to slips and trips, are reported each year. Many patients and visitors also receive injuries.

Recent evidence suggests that slips are also indirectly responsible for fatal accidents, as:

- slips are often the "first event" in falls from heights
- simple slip injuries (broken bones etc) often lead to complications in older people, such as thromboses or embolisms, which may be fatal.

Slips and trips risks are especially important in healthcare because:

- injuries to healthcare workers and members of the public are frequent
- trips account for almost 62 per cent of major injuries to members of the public. Generally trips are believed to account for between 25 per cent and 30 per cent of all slips, trips and falls
- they cause 8 per cent of fatalities to members of the public in the healthcare industry.

The National Audit Office, in their report of April 2003, highlighted slips and trips as a main type of accident to workers and patients. The report includes recommendations that NHS Trusts should review their health and safety risk management policies and improve accident reporting systems.

The cost of accidents could affect the delivery of high-quality patient care and viability of the business. For example:

- patients being seriously injured through falls leads to additional medical costs and an increased stay in hospital, with implications for waiting lists and service delivery
- staff sickness absence due to slips, trips and falls at work, and other associated costs, such as staff replacement costs, will have a detrimental effect on budgets
- the total estimated cost of civil claims for slips and trips injuries to employees and the public in or on NHS premises in England, reported to the NHS Litigation Authority over the past four years, exceeds £25 million. Typical claims average £5000, but some have been as high as £600 000. Increased insurance costs and enforcement of criminal liability can be a further consequence
- people may experience hardship through loss of wages, as well as pain and suffering.

RAILWAYS

The Rail Safety and Standards Board also produces statistics on passenger accidents and looks at the incidence of STFs. The data below is taken from the RSSB publication *The best flooring materials for stations – Phase 1*.

Accident statistics and contributory factors

Tables A3.1 and A3.2 list the numbers of passenger accidents at stations over recent years. Many accidents have been attributed to slips, trips and falls on level surfaces, and risk analysis has found that the risk contribution for slips, trips and falls represents over 50 per cent of the total risk for movement (involving trains) and non-movement accidents (Beaumont *et al*, 2004).

Table A3.2   *Passenger accidental fatalities: accidents at stations*

Where/how	Annual totals						
	2002/03	2001/02	2000/01	1999/00	1998/99	1997/98	1996/97
Boarding/alighting	2	1	1	1	1	3	0
Falls from platforms/ close to edge	4	3	2	4	9	6	6
Other	1	1	2	3	2	1	0
TOTAL	7	5	5	8	12	10	6

Table A3.3   *Passenger accidental major injuries: accidents boarding and alighting trains and accidents on railway premises unconnected with movement of trains*

Where/how	Annual totals						
	2002/03	2001/02	2000/01	1999/00	1998/99	1997/98	1996/97
Boarding/alighting	36	38	39	34	23	29	26
Slips/trips/falls	60	75	70	58	82	84	75
Using stairs	45	50	51	51	53	62	61
Using escalators	14	8	12	7	17	11	6
Using lifts etc	0	0	0	0	0	0	1
Falling over packages	6	4	7	0	13	6	7
Falling off platform	5	6	6	3	7	2	5
Electric shock	0	0	0	0	0	0	1
Other	4	4	10	13	18	22	6
TOTAL	170	185	195	166	213	216	188

Possible causal factors of accidents are manifold, but they include:

- design
- specification of materials and products
- installation
- material performance
- environmental conditions, eg wetness, ice, snow
- presence of contamination, litter, bird droppings, objects and obstacles
- cleaning and maintenance
- lighting (design and installation)
- background noise
- lack of warning signs and markings
- barriers
- handrails
- passenger behaviour
- congestion
- crowd control.

It is considered that whereas slips, trips and falls may occur anywhere, the probability of them occurring, and the risk of major injury, is greatest at certain locations and during certain activities, eg:

- at the edges of the platform when passengers are boarding and alighting from trains
- on escalators, particularly when joining and leaving
- on stairs and ramps
- where surfaces can be wetted or contaminated, eg areas exposed to the weather, areas near where food is sold or consumed, areas near where birds gather or roost, areas with litter
- where there is a risk of ice formation or snow during the winter

It should be noted than an accident is generally the result of a sequence of actions and events, and may well not have occurred had any of these developed differently. The complexity of the interaction complicates the study of accident causation, for subtle changes can have major consequences in the severity of accidents.

A4   Trips

A4.1   GENERAL

Trips occur when an obstruction prevents normal movement of the foot, resulting in a loss of balance. Usually caused by objects on the floor or due to uneven surfaces

(HSE, 1996a)

Figure A4.1   *Pallet: trip hazard (courtesy HSL)*

Slips and trips are often described together in discussions of the statistics of both occurrence and consequences (financial loss, injury, working days lost etc) (Loo-Morrey and Jeffries, 2003). The HSE "Watch your step" campaign is directed at both categories, and HSE maintains a "Slips and trips" website at <www.hse.gov.uk/slips/>. Both slips and trips may initiate falls, and guidance aimed at avoidance and statistics may cover all three.

HSE RIDDOR statistics indicate that trips account for 25–33 per cent of all STF accidents reported annually. This corresponds to between 8000 and 12 000 accidents each year (HSL, 2004).

The number of falls increases with age for both men and women, but the increase is more pronounced for the latter.

Despite the linking of slips and trips there are differences not only in their physical manifestation but also in the causative factors. These differences are important in dealing with the appropriate risk assessment and in mitigating any risk identified. Further research is planned by the HSE to enable the production of companion guidance to this work on slips.

Manning (1983) defined a "trip as the sudden arrest of movement of the foot with continued motion of the body".

For a trip to occur two conditions are usually required:

- a low obstacle in the pedestrian's path

- the pedestrian's inability or failure to see or notice the obstacle (WorkCover, 1998a).

ENVIRONMENTAL FACTORS

The HSE website <www.hse.gov.uk> reports that most trips result from poor housekeeping. This relates to factors such as trailing cables, uneven edges to flooring, gratings or covers, loose mats or carpet tiles, and temporary obstructions. Most of these are readily avoidable, but the need to address such hazards has to be a routine part of building usage and not something left for cleaning or maintenance staff – there are parallels, for example, with the obligation to report untied ladders and uncovered holes on construction sites. In terms of the Slip Potential Model described in Section 2.2, obstacles are the equivalent of "contamination" when considering trips and, like contamination, these are controllable.

Floors should be free of trip hazards, including ridges. It has been held in legal cases that a ridge as low as 10 mm can cause a trip. This needs to be considered in risk assessments. Low objects are more likely to lead to a trip, partly because they are less visible, but also because they may provoke instability and a subsequent fall. This is particularly true when the pedestrian is encumbered. Wires, cables and packaging are all trip hazards that fall into this category. Trip hazards also include changes of level, for example where flooring changes or there is a lowered matwell.

Figure A4.2   *Trailing cables (courtesy HSL)*

Externally, the same factors need to be considered. A common hazard in both private and public areas is loose and/or uneven paving. Similarly, changes in level between different surfaces, for example paved or tiled and asphalt, should be avoided. The area around tree pits is often uneven, either caused by near-surface root growth or settlement of the soil around the tree. This requires appropriate specification and control on installation, and subsequent maintenance as part of good management.

Figure A4.3   *Uneven paving (courtesy HSL)*

A simple testing device may be used when checking that paving has achieved the requirements of BS 7263-2 *Code of practice for laying precast concrete flags*. This standard requires that no two adjacent paving slabs shall have a difference in level greater than 3 mm.

A £1 coin, which is exactly 3 mm thick, when placed at the interface between two slabs can quickly be felt to detect a difference greater or less than 3 mm. Ideally the joints should be flush, but this test gives a clear and objective limit to acceptability and an unambiguous demonstration of unsatisfactory work.

Published guidance in the United Kingdom and overseas provides practical measures for trips risk control; a selection is included here, together with references where further advice of this kind is given.

Table A4.1   *Trip risk control (HSE, 2003d)*

Hazard	Suggested action
Trailing cables	Position equipment to avoid cables crossing pedestrian routes, use cable covers to securely fix to surfaces, restrict access to prevent contact. Consider use of cordless tools. Remember that contractors will also need to be managed
Rugs, mats	Ensure mats are securely fixed and do not have curling edges
Poor lighting	Improve lighting levels and placement of light fittings to ensure more even lighting of all floor areas

Trip hazards in construction

Trip hazards are common on construction sites, so it is important to maintain good standards of housekeeping and a high level of vigilance.

Source: British Cement Association

Preventing slips, trips and falls (Queensland Government DIR, nd) contains further advice on this topic:

- *offices*: provide power, telephone and computer services from ducts in the floor or from the ceiling, and fit out offices to provide flexibility in layout without requiring cords on the floor
- *factories*: support electric cords and pneumatic hoses for air tools overhead and keep them off the floor
- *construction*: avoid extension cords where possible by using battery-operated tools
- *other industries*: install additional power or telephone outlets to eliminate any cords on the floor.

It refers further to the need to avoid the uneven laying of paving slabs by laying them on a stable substrate, and suggests other controls to reduce tripping hazards as follows:

- mark aisles clearly
- provide sufficient storage systems to keep materials out of aisles
- use quite thin standing mats with tapered edges rather than traditional duckboards.

In addition to the above, changes in both types of flooring and in level need to be clearly delineated. This may be achieved by colour contrast, and good levels of illumination are also needed as noted in Table A4.1 above.

Figure A4.4   *Colour contrast strips indicate changes of level on a carpeted surface (courtesy HSE)*

A4.3   TOE CLEARANCE

As implied, toe clearance is a measure of the distance between the toe and the walking surface – or an obstacle on the surface. Research carried out by the HSL in 2003 (Loo-Morrey and Jeffries) included a review of data available regarding toe clearance among the healthy population. It concluded that this is "severely limited" and that this might be a suitable area for future research in terms of understanding how variation in this measure "with age and gender among the healthy population" may lead to a better understanding of trip accidents. The limited information available gives a high degree of scatter, from
8.7 mm to 21.9 mm for healthy young adults, with a suggested average of 14.5 mm. There is better agreement that the clearance decreases with increasing age, although the rate of decline is unclear.

Thus where there is a high percentage of elderly subjects in any given location or facility the need to avoid uneven surfaces is of particular importance.

Where it is not possible to avoid obstacles then precautions should be taken to reduce the risk of accident by preventing access or using warning signs or cones. Other studies have shown that where there is forewarning of a trip hazard, such as a uneven paving, subjects modify their gait significantly, with an increase in toe clearance of up to 50 per cent.

A5   Other techniques of surface measurement

> This appendix should be read in conjunction with Chapter 3.

Many commercially available measurement techniques exist for establishing the slipperiness of walking surfaces, via measurement of either the coefficient of friction or surface roughness. Some of these are described below for information but, as noted in Chapter 3, these are not specifically recommended by HSE. Although some tests have shown some promising correlation (Hallas *et al*, 2005) in respect of specific apparatus (mentioned below), they are considered by HSL to have limitations compared to the standard test methodology (the pendulum test and surface roughness measurement).

A5.1   GENERAL

HSE/HSL has indicated very clearly (HSE, 2004a) that current commercially available sled-type test instruments are not considered an acceptable means of measuring floor surface slipperiness. Although they may give comparable and reliable results in dry conditions, they rarely do so in surface-wet conditions. In these surface-wet cases, which is the situation for most slip occurrences, HSE believes that they fail to reproduce the correct squeeze film (Hallas *et al*, 2005). This failure routinely leads to inaccurate and misleading slip resistance values for wet contaminated floors.

HSE report (Hallas *et al*, 2005) that any valid test mechanism must reproduce the same type of squeeze film fluid dynamics as is seen during pedestrian slipping. To date the only portable test to do this is the pendulum as described in Section 2.4 and hence other sled tests must correlate with the Pendulum if they are to be useful.

A5.2   SLED TESTS

As noted above, HSE has found that laboratory-based assessments strongly suggest that several tests employing the sled principle can produce misleading data (HSL, 2002). Information from such tests also shows that some smooth floorings appear to be less slippery in wet conditions than when dry; this is clearly at odds with everyday experience (HSE, 2004a). HSE notes that sled tests are capable of producing accurate data when used to assess floors in clean, dry or dusty conditions, but points out that the majority of accidents occur in wet contaminated conditions.

Examples of sled tests include the German GMG100 test and the FSC2000 sled test. Both are described in HSL report PE/02/08 *Pedestrian slipping: slopes and encumbrance* (Lemon *et al*, 2002b). American slip resistance measurement is covered in ANSI A2164.2-2001 *Standard for the provision of slip resistance on walking/working surfaces*.

The Tortus Friction Tester is produced by Severn Science (Instruments) Ltd. In 1995 HSE stated (Rowland, 1995) that it was unable to accept the Tortus instrument or other instruments using the same principle of test. A detailed case for acceptance was made by the Tortus manufacturer in February 1998 (SSI, 1998), but soon afterwards it was accepted that the Tortus had significant shortcomings (TTA, 2002).

More recently HSL examined the Kirchberg and SlipAlert instruments (Hallas *et al*, 2005). Both instruments have shown an ability to distinguish between wet and dry floors. Although the narrower range of both the Kirchberg and SlipAlert can result in misclassification of dry floors, their main application will be in assessing wet contaminated surfaces (the conditions in which most slip accidents occur) and therefore the wet results are of more importance.

Of the two, SlipAlert appears to be able to generate a wider range of results and for this reason produces results that better reflect the pendulum SRV numbers. Where there is a difference between the CoF values generated and the pendulum SRV numbers, the test generally errs on the side of caution by underestimating the level of available slip resistance. It is therefore unlikely that an unsafe floor would be erroneously classified as safe by SlipAlert.

One factor that must be taken into consideration is the amount of room required to use a "roller coaster" mechanism such as SlipAlert compared with the pendulum. When the level of available friction is very low the "trolley" can travel more than 2 m. In many areas of site test there will not be sufficient room to use SlipAlert or Kirchberg correctly. The SlipAlert trolley is also prone to veer off to one side where there are areas of unevenness in the floor being tested, which can produce misleading results. In order for SlipAlert to be used as a standard test, four-S sliders would have to be adopted to allow reproducible conditions and results that could be compared with those obtained by the pendulum. When conditioned and operated in a standardised manner, SlipAlert presents a useful complementary tool to the surface roughness meter, and a possible alternative to the pendulum for basic slip resistance measurements in contaminated conditions. HSE/HSL consider SlipAlert to be a promising test method and is carrying out an in-depth investigation of its validity to compare it with the pendulum.

Technology is continually evolving; those researching suitable instruments for measuring surface slipperiness will need to ensure they are aware of the latest developments regarding instrument behaviour and suitability.

A5.3 SATRA FRICTION TESTER*

SATRA TM 144:2004 utilises a slip rig to measure slip resistance in the laboratory. Although it is developed for footwear testing the test is nonetheless based on biomechanical studies of slips. The test defines a dynamic coefficient of friction at a constant speed under a vertical load representative of a human body weight. The test slider generally uses 96 rubber (ie four-S).

The test allows for an assessment of differing footwear, sole tread, pattern and surface contamination to give realistic performance values under normal service conditions. It is thought to give useful information specifically on resilient and profiled floors and HSL is exploring the possibility of it becoming a test that may be used for flooring.

A5.4 GERMAN DIN RAMP TEST

It was from this test, used by many European flooring manufacturers to classify the slipperiness of their products before sale, that the modified HSL ramp test was developed (described in Section 3.4).

---

* The text is based on Ferry (2005).

The test follows either DIN 51097 (which utilises barefoot operators with a soap solution contaminant) or DIN 51130 (which uses heavily cleated EN345 safety boots with motor oil as a contaminant). The HSL ramp test utilises water.

In both cases the test subject walks forwards and backwards over the contaminated flooring sample. The inclination of the ramp is gradually increased until the test subject slips. The average angle of inclination at which slip occurs can be used to calculate the coefficient of friction of the level flooring.

The classification scheme outlined in the DIN standards has led to some confusion and misapplication of floor surfaces in the UK (HSE, 2004a).

HSE reports that a common misconception is that the "R" scale runs from R1 to R13, in decreasing slipperiness. In fact, the scale runs from R9–R13, in decreasing slipperiness, as noted in Table A5.1. Thus R9 is not at the mid-point, but is at the lowest (most slippery) end of the scale. Floor surfaces classified as R9, or in some circumstances R10, are likely to be unacceptably slippery in wet or greasy conditions.

Table A5.1 *DIN "R" slipperiness values (Table 4 from DIN 51130 "R value" slipperiness classification regime)*

Classification	R9	R10	R11	R12	R13
Slip angle (degrees)	3–10	10–19	19–27	27–35	>35

The slip angles are equivalent to the following coefficients of friction:

Classification	R9	R10	R11	R12	R13
Coefficient of friction	0.05–0.18	0.18–0.34	0.34–0.51	0.51–0.70	>0.70

A6 References

A6.1 PUBLICATIONS

Articles and publications marked with an asterisk (*) are available as free web downloads. Where known, the web address for the download is given.

AT&T (2005)
* *Avoiding slips, trips and falls in the office*
Available from: <www.att.com/ehs/safety/slips_trips_falls.html>

AUSTRALIAN INSTITUTE OF ENVIRONMENTAL HEALTH (1993)
National code for the construction and fitout of food premises
Australian Institute of Environmental Health, Canberra

BAILEY, M (2005)
"The TRL pendulum slip resistance tester. The reasoning for its acceptance in the UK"
Contemporary Ergonomics, pp 493–497

BEAUMONT, R J, BURTWELL, M H and JORDAN, R W (2004)
* *The best flooring materials for stations – Phase 1. Literature review on reducing slips and other hazards on flooring materials for stations*
T157a, PR.CPS/048/03, Rail Standards Safety Board, London
Available from: <www.rssb.co.uk>

BLR (2005)
* *Slips, trips and fall hazards are everywhere – preventing them is easy*
Business and Legal Reports, Old Saybrook
Available at: <www.blr.com/product.cfm/product/10008900/funcode/>

BOWMAN, R (nd)
* *Where to next with slip resistance standards?*
CSIRO, Highett
Available at: <www.cmit.csiro.au/vb2/345/67/>

BRE (2001)
The effect of going size on stair safety
80503, Building Research Establishment, Garston

BRE (2002a)
The effect of use of different materials on stair slipping accidents
Client Report 203492, Building Research Establishment, Garston

BRE (2002b)
Use of proprietary nosing in an overstep situation
Client Report 203493, Building Research Establishment, Garston

BROUGHTON, R A; ROWLAND, F J; GRIFFITHS, R S and RICHARDSON, M T (1997)
Research into aspects of the pedestrian slipping problem – Phase II
IR/L/PE/97/6, Health and Safety Laboratory, Buxton

BUNTERNGCHIT, Y; LOCKHART, T; WOLDSTAD, J C and SMITH, J L (2000)
"Age related effects of transitional floor surfaces and obstructions of view on gait characteristics related to slips and falls"
International Journal of Industrial Ergonomics, vol 25, pp 223–232

CIBSE (2002)
Code for lighting (with 2004 addendum)
Butterworth-Heinemann, Oxford (ISBN 0-7506-5637-9)

COMCARE (2002)
* *Guide to preventing, slips, trips and falls*
OHS 34, Safety, Rehabilitation and Compensation Commission, Canberra
Available at: <www.comcare.gov.au/pdf_files/OHS_34_slips_trips_falls_quick_ref_guide_jun04_v1.pdf>

COUNTY SURVEYOR'S SOCIETY (1996)
The assessment of slip resistance in paved areas used by pedestrians and horse riders
Eng/1-96, County Surveyor's Society, London

COX, S J and O'SULLIVAN, E F (1995)
Building regulation and safety
BR 290, Building Research Establishment, Garston

DEPARTMENT FOR TRANSPORT (2002)
* *Inclusive mobility – a guide to best practice on access to pedestrian and transport infrastructure*
Department for Transport, London
Available at: <www.dft.gov.uk/stellent/groups/dft_mobility/mobility/documents/pdf/dft_mobility_pdf_503282.pdf>

DEPARTMENT OF TRADE AND INDUSTRY (1999)
* *Avoiding slips, trips and broken hips. Guidance for professionals who work with older people*
Department of Trade and Industry, London
Available at: <www.dti.gov.uk/homesafetynetwork/pdffalls/guidance.pdf>

DEPARTMENT OF TRADE AND INDUSTRY (2000)
* *Step up to safety. Information for older people on how to use the stairs safely*
URN 00/537, Department of Trade and Industry, London (ISBN 0-7521-1914-1)
Available at: <www.dti.gov.uk/homesafetynetwork/pdffalls/booklet.pdf>

DEPARTMENT OF THE ENVIRONMENT, TRANSPORT AND THE REGIONS (1999)
* *Guidance on the use of tactile paving surfaces*
Department of the Environment, Transport and the Regions, London
Available at: <www.dft.gov.uk/stellent/groups/dft_mobility/documents/pdf/dft_mobility_pdf_503283.pdf>

EASTERBROOK, L; HORTON, K; ARBER, S and DAVIDSON, K (2001)
International review of interventions in falls among older people
URN/01/1173, Health Development Agency, London

EUROPEAN CLEANING JOURNAL (1997)
"Robotics – will you be a pioneer?"
European Cleaning Journal, Jun/Jul

FERRY, S (2005)
"Slip testing of floors"
SATRA Spotlight, Oct, p 6

FONG, D T P; HONG, Y and LI, J X (2004)
"Changes of gait pattern on slippery walking surfaces in simulated construction worksite environments"
In: *Proc XXIInd int symp biomechanics in sports*, Ottawa, Canada, pp 538–540

GIBSON, I (2005)
Guidance to prevent slips, trips and falls
British Cement Association, Camberley

HALLAS, K; SHAW, R; LEMON, P and THORPE, S (2005)
"Roller coaster slip tests: putting slip testing back on the rails"
Contemporary Ergonomics, pp 514–518

HARDING, J R and SMITH, R A (1995)
Rigid paving with clay pavers
Design Note 8, Brick Development Association, Winkfield

HEALTH AND SAFETY COMMISSION (2000)
Revitalising health and safety. Strategy statement
OSCSGO390, Department of Environment, Transport and the Regions, London

HEALTH AND SAFETY COMMISSION (2001)
* *Health and safety statistics 2000/01*
HSE Books, Sudbury
Available from: <www.hse.gov.uk/statistics/pdf/hss0001.pdf>

HEALTH AND SAFETY COMMISSION (2004)
* *Statistics of workplace fatalities and injuries. Slips and trips*
HSE Books, Sudbury
Available from: <www.hse.gov.uk/statistics/pdf/rhsslips.pdf>

HEALTH AND SAFETY EXECUTIVE (1996a)
Slips and trips. Guidance for employers on identifying hazards and controlling risks
HSG155, HSE Books, Sudbury (ISBN 0-7176-1145-0)

HEALTH AND SAFETY EXECUTIVE (1996b)
Slips and trips. Guidance for the food processing industry
HSG156, HSE Books, Sudbury (ISBN 0-7176-0832-8)

HEALTH AND SAFETY EXECUTIVE (1996c)
* *Slips and trips: summary guidance for the catering industry*
FIS6, HSE Books, Sudbury
Available from: <www.hse.gov.uk/pubns/fis06.pdf>

HEALTH AND SAFETY EXECUTIVE (1996d)
Health and safety in construction
HSG150, HSE Books, Sudbury (ISBN 0-7176-2106-5)

HEALTH AND SAFETY EXECUTIVE (1998)
Lighting at work
HSG38, HSE Books, Sudbury (ISBN 0-7176-1232-5)

HEALTH AND SAFETY EXECUTIVE (1999)
* *Preventing slips in the food and drink industries – technical update on floor specifications*
FIS22, HSE Books, Sudbury
Available from: <www.hse.gov.uk/pubns/fis22.pdf>

HEALTH AND SAFETY EXECUTIVE (2001)
Managing health and safety in construction. Approved code of practice and guidance
HSG224, HSE Books, Sudbury (ISBN 0-7176-2139-1)

HEALTH AND SAFETY EXECUTIVE (2003a)
* *Slips and trips speaker's pack*
HSE Books, Sudbury
Available from: <www.hse.gov.uk/slips/information.htm>

HEALTH AND SAFETY EXECUTIVE (2003b)
* *Preventing slip and trip incidents in the education sector*
EDIS2, HSE Books, Sudbury
Available from: <www.hse.gov.uk/pubns/edis2.pdf>

HEALTH AND SAFETY EXECUTIVE (2003c)
* *Slips and trips in the health service*
HSIS2, HSE Books, Sudbury
Available from: <www.hse.gov.uk/pubns/hsis2.pdf>

HEALTH AND SAFETY EXECUTIVE (2003d)
* *Preventing slips and trips at work*
INDG225(rev1), HSE Books, Sudbury (ISBN 0-7176-2760-8)
Available from: <www.hse.gov.uk/pubns/indg225.pdf>

HEALTH AND SAFETY EXECUTIVE (2004a)
* *The assessment of pedestrian slip risk. The HSE approach*
SAT1, HSE Books, Sudbury
Available from: <www.hse.gov.uk/pubns/web/slips01.pdf>

HEALTH AND SAFETY EXECUTIVE (2004b)
* *The slips assessment tool*
HSE Books, Sudbury
Available from: <www.hsesat.info>

HEALTH AND SAFETY EXECUTIVE (2005a)
* *Slips and trips: the importance of floor cleaning*
SAT2, HSE Books, Sudbury
Available from: <www.hse.gov.uk/pubns/web/slips02.pdf>

HEALTH AND SAFETY EXECUTIVE (2005b)
* *Preventing slips and trips in kitchens and food service*
CAIS6, HSE Books, Sudbury
Available from: <www.hse.gov.uk/pubns/cais6.pdf>

HEALTH AND SAFETY LABORATORY (1994)
Research into the slip resistance of floors. Site visits
IR/L/PE/94/5, Health and Safety Laboratory, Buxton

HEALTH AND SAFETY LABORATORY (2000)
Assessment of slip potential: British Sugar plc
PE/00/13, Health and Safety Laboratory, Buxton

HEALTH AND SAFETY LABORATORY (2003)
Slip and trip human factors scoping study – development of safety performance measures
ERG/03/06, Health and Safety Laboratory, Buxton

HEALTH AND SAFETY LABORATORY (2004)
Review of RIDDOR trip accident statistics 1991–2001
PS/02/04, Health and Safety Laboratory, Buxton

HELA (2001)
National picture 2001
HSE/LA Enforcement Liaison Committee, Health and Safety Commission, London

HIGHWAYS AGENCY (2004)
*Design manual for roads and bridges. Vol 2 Highway structures: design (substructures). Materials.
Section 2 Special structures. Part BD 29/04 Design criteria for footbridges*
The Stationery Office, London
Available from: <www.archive2.official-documents.co.uk/document/deps/ha/dmrb/
vol2/sect2/bd2904.pdf>

HOLAH, J (1994)
"Hygiene and safety in the food industry: compromise or complimentary" [sic]
In: *Slipping – towards safer flooring*, RAPRA, Shawbury, 29 Sep 1994 (seminar)
Campden and Chorleywood Food Research Association, Chipping Campden
(ISBN 1-85957-025-9)

HUGHES, R C and JAMES, D I (1994)
"Slipping determinations – magic and myth"
In: *Slipping – towards safer flooring*, RAPRA, Shawbury, 29 Sep 1994 (seminar)
Campden and Chorleywood Food Research Association, Chipping Campden (ISBN 1-85957-025-9)

HYDE, A S; BAKKEN, G M; ABELE, J R; COHEN, H H and LaRUE, C A (2002)
Falls and related injuries: slips, trips, missteps and their consequences
Lawyers and Judges Publishing Company, Tucson

IDDON, J and CARPENTER, J (2004)
Safe access for maintenance and repair
C611, CIRIA, London

ISOES (2005)
Proceedings
International Society for Occupational Ergonomics and Safety, Las Vegas

KACZMAR, P M (2001)
Sealing timber floors. A best practice guide to floor preparation and the selection, application and maintenance of floor lacquers
Report 2/2001, TRADA Technology, High Wycombe (ISBN 1-90051-033-2)

KACZMAR, P M (2002)
Seals for timber floors: a specification guide
Report 2/2002, TRADA Technology, High Wycombe (ISBN 1-90051-036-7)

KEIL CENTRE (2000)
* *Behaviour modification to improve safety: literature review*
OTO2000/003, HSE Books, Sudbury (ISBN 0-7176-1893-5)
Available from: <www.hse.gov.uk/research/otopdf/2000/oto00003.pdf>

KLOSS, G (nd)
Trips, slips and falls in SMEs – analysing risks in the meat industry
Berufsgenossenschaftliches Institut für Arbeitssicherheit, quoted in *Safety Science Monitor*, vol 7 no 4

LEHTOLA, C J; BECKER, W J and BROWN, C M (2001)
Preventing injuries from slips trips and falls
National Ag Safety Database CIR869, Institute of Food and Agricultural Sciences, University of Florida, Gainesville

LEMON, P W (2003)
Pedestrian safety: a study of the slipperiness of real workplace floor surface contamination
PS/03/03, Health and Safety Laboratory, Buxton

LEMON, P W and GRIFFITHS, R S (1997)
Further application of squeeze film theory to pedestrian slipping
IR/L/PE/97/9, Health and Safety Laboratory, Buxton

LEMON, P W and ROWLAND, F J (1997)
Pedestrian slipping: the risk presented by inappropriate shoe/floor material combinations
IR/L/PE/97/12, Health and Safety Laboratory, Buxton

LEMON, P W; ROWLAND, F J; THORPE, S and HUGHES, A (nd)
Assessment of the slip resistance of grating panels with the DIN ramp test
Health and Safety Laboratory, Buxton

LEMON, P W; THORPE, S C and GRIFFITHS, R S (1999a)
Pedestrian slipping Phase 4: macro-rough and profiled floors
IR/L/PE/99/01, Health and Safety Laboratory, Buxton

LEMON, P W; THORPE, S C and GRIFFITHS, R S (1999b)
Pedestrian slipping – cleaning and surface treatment
IR/L/PE/99/05, Health and Safety Laboratory, Buxton

LEMON, P W; THORPE, S C; JEFFRIES, S L; SEXTON, C and HAWKINS, M (2001)
Pedestrian slipping: dry contaminants
PE/01/15, Health and Safety Laboratory, Buxton

LEMON, P W; THORPE, S C; JEFFRIES, S; HAWKINS, M; LOO-MORREY, M and
BROWN, E (2002a)
Pedestrian slipping: overshoes
PED/04/01, Health and Safety Laboratory, Buxton

LEMON, P W; THORPE, S C; JEFFRIES, S; HAWKINS, M and SEXTON, C (2002b)
Pedestrian slipping: slopes and encumbrance
PE/02/08, Health and Safety Laboratory, Buxton

LONDON UNDERGROUND LIMITED (2005a)
Premises – finishes
Standard 2-01107-007, Issue A1, London Underground Limited, London

LONDON UNDERGROUND LIMITED (2005b)
Premises – stairways and ramps
Standard 2-01107-005, Issue A1, London Underground Limited, London

LOO-MORREY, M (2005a)
Slip testing of occupational footwear
PED/05/04, Health and Safety Laboratory, Buxton

LOO-MORREY, M (2005b)
Tactile paving survey
PED/05/03, Report 2005/07, Health and Safety Laboratory, Buxton

LOO-MORREY, M and HALLAS, K (2003)
Evaluation of SBR rubber for use as slider in pendulum test
PS/RE/13/2003, Health and Safety Laboratory, Buxton

LOO-MORREY, M; HALLAS, K A and THORPE, S C (2004)
"Minimum going requirements and friction demands during stair descent"
Contemporary Ergonomics, pp 8–12

LOO-MORREY, M and JEFFRIES, S L (2003)
Trip feasibility study
PS/03/05, Health and Safety Laboratory, Buxton

LOUGHBOROUGH UNIVERSITY and UMIST (2003)
* *Causal factors in construction accidents*
RR156, HSE Books, Sudbury (ISBN 0-7176-2749-7)
Available from: <www.hse.gov.uk/research/rrpdf/rr156.pdf>

MANNING, D P (1983)
"Deaths and injuries caused by slipping, tripping and falling"
Ergonomics, vol 26, no 1, pp 3–9

MASON, S (2003)
* *Development of a methodology for the assessment of human factors issues relative to trips, slips
and fall accidents in the offshore industries*
RR065, HSE Books, Sudbury (ISBN 0-7176-2168-5)
Available from: <www.hse.gov.uk/research/rrpdf/rr065.pdf>

MAYNARD, W S (2002)
"Tribology: preventing slips and falls in the workplace"
* *Liberty Directions*, Spring, pp 13–17
Available from: <www.libertymutual.com>

MILLER, B C (1998)
* *Investigating slips and falls: the complex dynamics behind simple accidents*
Available from: <www.safety-engineer.com/complex.htm>

MURRAY, M P (1967)
"Gait as a total pattern of movement"
American Journal of Physical Medicine, vol 46, pp 290–333

MURRAY, M P; KORY, R C and CLARKSON, B H (1969)
"Walking patterns in healthy old men"
Journal of Gerontology, vol 24, pp 169–178

NAGATA, H (1991)
"Analysis of fatal falls on the same level or on steps/stairs"
Safety Science, vol 14, pp 213–222

NATIONAL FLOOR SAFETY INSTITUTE (2002)
* "Slip and fall accidents no joke"
Inside Edition, 8 Nov
Available from: <http://consumeraffairs.com/news02/slipfall.html>

NATIONAL FLOOR SAFETY INSTITUTE (2003)
Restaurant slip-and-fall accident prevention programme
National Floor Safety Institute, Southlake

NEWTON, R A (1997)
* *The fall prevention program manual*
Department of Physical Therapy, College of Allied Health Professions, Temple
University, Philadelphia
Available from: < www.temple.edu/older_adult/fppmanual.html>

NHS ESTATES (1995)
Flooring, 2nd edn
Building Components, HTM 61, HMSO, London (ISBN 0-11-322203-3)

NHS ESTATES (2004)
* *The NHS healthcare cleaning manual*
Department of Health, London
Available from: <patientexperience.nhsestates.gov.uk/clean_hospitals/ch_content/
cleaning_manual/background.asp>

PEEBLES, L; WEARING, S and HEASMAN, T (2004)
* *Identifying human factors associated with slip and trip accidents*
RR382, HSE Books, Sudbury (ISBN 0-7176-6160-1)
Available from: <www.hse.gov.uk/research/rrpdf/rr382.pdf>

PYE, P W and HARRISON, H W (2003)
*Floors and flooring: performance, diagnostics, maintenance and, repair and the avoidance of
defects*, 2nd edn
Building Research Establishment, Garston (ISBN 1-8608-1631-2)

QUEENSLAND GOVERNMENT DEPARTMENT OF INDUSTRIAL RELATIONS (nd)
Slips, trips and falls – risk management tool
State of Queensland Depart of Industrial Relations, Brisbane

RAIL SAFETY AND STANDARDS BOARD (2005)
Annual safety performance report 2004
Rail Safety and Standards Board, London

REASON, J T (1997)
Managing the risks of organizational accidents
Ashgate Publishing, Aldershot (ISBNs 1-84014-104-2, 1-84014-105-0)

ROSSMORE GROUP (2003)
Slips, trips and falls. Addressing the "quiet" epidemic in today's railway
Rossmore Group, Solihull

ROSSMORE GROUP (2005)
Research into the behavioural aspects of slip and trip accidents and incidents. Pt 1 Literature review
HSE Books, Sudbury

ROSSMORE GROUP (nd)
Railway STF workshops
Rossmore Group, Solihull

ROWLAND, F J (1995)
Solicitors' statement on behalf of HSE
Health and Safety Laboratory, Buxton

ROYS, M S (2001)
Safe as houses?
DG 458, Building Research Establishment, Garston (ISBN 1-8608-1499-9)

ROYS, M S and WRIGHT, M (2003)
Proprietary nosings for non-domestic stairs
IP 15/03, Building Research Establishment, Garston (ISBN 1-8608-1652-5)

SENTINELLA, J; WELLS, P and FLOWER, C (2005)
* *The use of tactile surfaces at rail stations: final report*
T158, Rail Safety and Standards Board, London
Available from: <www.rssb.co.uk>

SEVERN SCIENCE (INSTRUMENTS) LTD (1998)
The Tortus floor friction tester – the case for use in either dry, wet or contaminated conditions
Submitted to Health and Safety Laboratory, 6 Feb

SHAW, R; HALLAS, K; LEMON, P and THORPE, S (2004)
Floor cleaning – drop that mop!
Health and Safety Laboratory, Buxton

STRATEGIC RAIL AUTHORITY (2005)
* *Train and station services for disabled passengers. A code of practice*
Strategic Rail Authority, London
Available from: <www.sra.gov.uk/pubs2/consultations/> (see "Railways for all")

TAYLOR, S N (1998)
Workplace floor surfaces. Safety issues
SIR55, HSE Books, Sudbury (ISBN 0-7176-2801-9)

THORPE, S C and LEMON, P W (2000)
"Pedestrian slipping – a risk assessment based approach"
In: *Proc IEA 2000/HFES 2000 congress*, San Diego
International Ergonomics Association/Human Factors Ergonomics Society, vol 4,
pp 486–489

THORPE, S C and LOO-MORREY, M (2003)
Study of the dynamics of pedestrian stair use
PS/03/02, Health and Safety Laboratory, Buxton

THE TILE ASSOCIATION (2002)
Slip resistance of hard flooring
The Tile Association, Beckenham

THE TILE ASSOCIATION (2004)
* *The design and installation of wall and floor tiling in food preparation, treatment and processing areas*
The Tile Association, Beckenham
Available from: <www.tiles.org.uk/help/answer-food.pdf>

THE TILE ASSOCIATION (nd)
The cleaning of ceramic tiles
The Tile Association, Beckenham

TIMPERLEY, D A (2002)
Guidelines for the design and construction of floors for food production areas, 2nd edn
Guideline no 40, Campden and Chorleywood Food Research Association, Chipping Campden

TRADA TECHNOLOGY (1999)
Timber decking manual
DG-2, Trada Technology, High Wycombe (ISBN 1-90051-022-7)

TURNER, B A; PIDGEON, N; BLOCKLEY, D I and TOFT, D B (1989)
Safety culture – its importance in future risk management
Position paper for 2nd World Bank workshop on safety control and risk management, Sweden

UCL (2005)
Personal communication

UK SLIP RESISTANCE GROUP (2000)
The measurement of floor slip resistance. Guidelines recommended by the UK Slip Resistance Group
Issue 2, Rapra Technology, Shawbury

UK SLIP RESISTANCE GROUP (2005)
The measurement of floor slip resistance. Guidelines recommended by the UK Slip Resistance Group
Issue 3, Rapra Technology, Shawbury

UNISON (2004)
* *Health and safety information sheet: slips and trips*
Unison, London
Available at: <www.unison.org.uk/acrobat/B1213.pdf>, accessed 1 Dec 2005

WORKCOVER AUTHORITY OF NEW SOUTH WALES (1998a)
Preventing slips, trips and falls – guidance note
Catalogue No 743, WorkCover Authority of New South Wales, Gosford

WORKCOVER AUTHORITY OF NEW SOUTH WALES (1998b)
Health and safety for cleaners: selecting the right mopping equipment
Catalogue No 751, WorkCover Authority of New South Wales, Gosford

YATES, T J S and RICHARDSON, D (2000)
Flooring, paving and setts: requirements for safety in use
IP 10/00, Building Research Establishment, Garston (ISBN 1-8608-1372-0)

A6.2 HSE PUBLICATIONS (BY CODE)

CAIS6 *Preventing slips and trips in kitchens and food service* (HSE, 2005b)

EDIS2 *Preventing slip and trip incidents in the education sector* (HSE, 2003b)

FIS6 *Slips and trips: summary guidance for the catering industry* (HSE, 1996c)

FIS22 *Preventing slips in the food and drink industries – technical update on floor specifications* (HSE, 1999)

HSG38 *Lighting at work* (HSE, 1998)

HSG150 *Health and safety in construction* (HSE, 1996d)

HSG155 *Slips and trips. Guidance for employers on identifying hazards and controlling risks* (HSE, 1996a)

HSG156 *Slips and trips. Guidance for the food processing industry* (HSE, 1996b)

HSG224 *Managing health and safety in construction. Approved code of practice and guidance* (HSE, 2001)

HSIS2 *Slips and trips in the health service* (HSE, 2003c)

INDG225(rev1) *Preventing slips and trips at work* (HSE, 2003d)

OTO2000/003 *Behaviour modification to improve safety: literature review* (Keil Centre, 2000)

RR065 *Development of a methodology for the assessment of human factors issues relative to trips, slips and fall accidents in the offshore industries* (Mason, 2003)

RR382 *Identifying human factors associated with slip and trip accidents* (Peebles, Wearing and Heasman, 2004)

SAT1 *The assessment of pedestrian slip risk. The HSE approach* (HSE, 2004a)

SAT2 *Slips and trips: the importance of floor cleaning* (HSE, 2005a)

SIR55 *Workplace floor surfaces. Safety issues* (Taylor, 1998)

A6.3 HSL PUBLICATIONS (BY CODE)

ERG/03/06 *Slip and trip human factors scoping study – development of safety performance measures* (HSL, 2003)

IR/L/PE/94/5 *Research into the slip resistance of floors. Site visits* (HSL, 1994)

IR/L/PE/97/6 *Research into aspects of the pedestrian slipping problem – Phase II* (Broughton *et al*, 1997)

IR/L/PE/97/9 *Further application of squeeze film theory to pedestrian slipping* (Lemon and Griffiths, 1997)

IR/L/PE/97/12 *Pedestrian slipping: the risk presented by inappropriate shoe/floor material combinations* (Lemon and Rowland, 1997)

IR/L/PE/99/01 *Pedestrian slipping Phase 4: macro-rough and profiled floors* (Lemon, Thorpe and Griffiths, 1999a)

IR/L/PE/99/05 *Pedestrian slipping – cleaning and surface treatment* (Lemon, Thorpe and Griffiths, 1999b)

PE/00/13 *Assessment of slip potential: British Sugar plc* (HSL, 2000)

PE/01/15 *Pedestrian slipping: dry contaminants* (Lemon *et al*, 2001)

PE/02/08 *Pedestrian slipping: slopes and encumbrance* (Lemon *et al*, 2002b)

PED/04/01 *Pedestrian slipping: overshoes* (Lemon *et al*, 2002)

PED/05/03 *Tactile paving survey* (Loo-Morrey, 2005b)

PED/05/04 *Slip testing of occupational footwear* (Loo-Morrey, 2005a)

PS/02/04 *Review of RIDDOR trip accident statistics 1991–2001* (HSL, 2004)

PS/03/02 *Study of the dynamics of pedestrian stair use* (Thorpe and Loo-Morrey, 2003)

PS/03/03 *Pedestrian safety: a study of the slipperiness of real workplace floor surface contamination* (Lemon, 2003)

PS/03/05 *Trip feasibility study* (Loo-Morrey and Jeffries, 2003)

PS/RE/13/2003 *Evaluation of SBR rubber for use as slider in pendulum test* (Loo-Morrey and Hallas, 2003)

A6.4 STANDARDS

A6.4.1 BS

BS 812-114:1989. *Testing aggregates – method for determination of the polished-stone value* (superseded)

BS 1134-1:1988. *Assessment of surface texture. Part 1: Methods and instrumentation*

BS 4211:1994. *Specification for ladders for permanent access to chimneys, other high structures, silos and bins* (superseded by 2005 edition)

BS 4592-1:1995. *Industrial type metal flooring, walkways and stair treads specification for open bar gratings*

BS 4592-2:1987. *Industrial type flooring, walkways and stair treads specification for expanded metal grating panels*

BS 5325:2001. *Installation of textile floor coverings. Code of practice*

BS 5385-3:1989. *Wall and floor tiling. Part 3: Code of practice for the design and installation of ceramic floor tiles and mosaics*

BS 5385-5:1994. *Wall and floor tiling. Part 5: Code of practice for the design and installation of terrazzo tile and slab, natural stone and composition block floorings*

BS 5395-1:2000. *Stairs, ladders and walkways. Part 1: Code of practice for the design, construction and maintenance of straight stairs and winders*

BS 5395-3:1985. *Stairs, ladders and walkways. Code of practice for the design of industrial type stairs, permanent ladders and walkways*

BS 5588-5:1991. *Fire precautions in the design, construction and use of buildings. Code of practice for firefighting stairs and lifts* (superseded by 2004 edition)

BS 6263-2:1991. *Care and maintenance of floor surfaces. Part 2:Code of practice for resilient sheet and tile flooring*

BS 6431-1:1983. *Ceramic floor and wall tiles specification for classification and marking, including definitions and characteristics*

BS 6717:2001. *Precast, unreinforced concrete paving blocks. Requirements and test methods*

BS 7263-1:2001. *Precast concrete flags, kerbs, channels, edgings and quadrants. Part 1: Precast, unreinforced concrete paving flags and complementary fittings – requirements and test methods*

BS 7923:2003. *Determination of the unpolished and polished pendulum test value of surfacing units*

BS 7953:1999. *Entrance flooring systems. Selection, installation and maintenance*

BS 7976-1:2002. *Pendulum testers. Part 1: Specification*

BS 7976-2:2002. *Pendulum testers. Part 2: Method of operation*

BS 7976-3:2002. *Pendulum testers. Part 3: Method of calibration*

BS 7997:2003. *Products for tactile paving surface indicators – specification*

BS 8201:1987. *Code of practice for flooring of timber, timber products and wood based panel products* (formerly CP 201)

BS 8203:2001. *Code of practice for Installation of resilient floor coverings*

BS 8204-1:2003. *Screeds, bases and in situ floorings. Part 1: Concrete bases and cement sand levelling screeds to receive floorings – code of practice*

BS 8204-2:2003. *Screeds, bases and in-situ floorings. Part 2: Concrete wearing surfaces – code of practice*

BS 8204-3:2003. *Screeds, bases and in-situ floorings. Part 3: Polymer modified cementitious levelling screeds and wearing screeds – code of practice*

BS 8204-4:2004. *Screeds, bases and in situ floorings. Part 4: Cementitious terrazzo wearing surfaces –code of practice*

BS 8204-5:2004. *Screeds, bases and in situ floorings. Part 5: Mastic asphalt underlays and wearing surfaces – code of practice*

BS 8204-6:2001. *Screeds, bases and in situ floorings. Part 6: Synthetic resin floorings – code of practice*

BS 8204-7:2003. *Screeds, bases and in situ floorings. Part 7. Pumpable self-smoothing screeds – code of practice*

BS 8300:2001. *Design of buildings and their approaches to meet the needs of disabled people – code of practice*

A6.4.2 BS EN

BS EN 344:1993. *Safety protective and occupational footwear for professional use requirements and test methods*

BS EN 345:1997. *Safety footwear for professional use additional specifications*

BS EN 346:1993. *Protective footwear for professional use*

BS EN 548:2004. *Resilient floor coverings. Specification for plain and decorative linoleum*

BS EN 649:1997. *Resilient floor coverings. Homogeneous and heterogeneous polyvinyl chloride floor coverings – specification*

BS EN 650:1997. *Resilient floor coverings. Polyvinyl chloride floor coverings on jute backing or on polyester felt backing or on polyester felt with polyvinyl chloride backing – specification*

BS EN 651:1997. *Resilient floor coverings. Polyvinyl chloride floor coverings with foam layer – specification*

BS EN 652:1997. *Resilient floor coverings. Polyvinyl chloride floor coverings with cork-based backing – specification*

BS EN 653:1997. *Resilient floor coverings. Expanded (cushioned) polyvinyl chloride floor coverings – specification*

BS EN 654:1997. *Resilient floor coverings. Semi-flexible polyvinyl chloride tiles – specification*

BS EN 655:1997. *Resilient floor coverings. Tiles of agglomerated composition cork with polyvinyl chloride wear layer – specification*

BS EN 686:1997. *Resilient floor coverings. Specification for plain and decorative linoleum on a foam backing*

BS EN 687:1997. *Resilient floor coverings. Specification for plain and decorative linoleum on a corkment backing*

BS EN 688:1997. *Resilient floor coverings. Specification for corklinoleum*

BS EN 1097-8:2000. *Tests for mechanical and physical properties of aggregates. Part 8: Determination of polished stone value*

BS EN 1338:2003. *Concrete paving blocks. Requirements and test methods*

BS EN 1339:2003. *Concrete paving flags. Requirements and test methods*

BS EN 1341:2001. *Slabs of natural stone for external paving. Requirements and test methods*

BS EN 1342:2001. *Setts of natural stone for external paving. Requirements and test methods*

BS EN 1344:2002. *Clay pavers. Requirements and test methods*

BS EN 1816:1998. *Resilient floor coverings. Specification for homogeneous and heterogeneous smooth rubber floor coverings with foam backing*

BS EN 1817:1998. *Resilient floor coverings. Specification for homogeneous and heterogeneous smooth rubber floor coverings*

BS EN 12004:2001. *Adhesives for tiles. Definitions and specifications*

BS EN 12104:2000. *Resilient floor coverings. Cork floor tiles – specification*

BS EN 12199:1998. *Resilient floor coverings. Specifications for homogeneous and heterogeneous relief rubber floor coverings*

BS EN 13036-4:2003. *Road and airfield surface characteristics – test methods. Part 4: Method for measurement of slip/skid resistance of a surface – the pendulum test*

BS EN 13748-1:2004. *Terrazzo tiles. Part 1: Terrazzo tiles for internal use*

BS EN 13888:2002. *Grouts for tiles. Definitions and specifications*

BS EN 14411:2003. *Ceramic tiles. Definitions, classification, characteristics and marking*

A6.4.3   BS EN ISO

BS EN ISO 14122-1:2001. *Safety of machinery – permanent means of access to machinery choice of a fixed means of access between two levels*

BS EN ISO 14122-2:2001. *Safety of machinery – permanent means of access to machinery, working platforms and walkways*

BS EN ISO 14122-3:2001. *Safety of machinery – permanent means of access to machinery stairways, stepladders and guard-rails*

BS EN ISO 14122-4:2001. *Safety of machinery – permanent means of access to machinery stairways, fixed ladders*

A6.4.4   DD ENV

DD ENV 12633:2003. *Method of determination of unpolished and polished slip/skid resistance value*

A6.4.5　　　*DIN*

DIN 51097:1992. *Testing of floor coverings: determination of the anti- slip properties; wet-loaded barefoot areas; walking method; ramp test*

DIN 51130:2004. *Testing of floor coverings; determination of the anti-slip properties; workrooms and fields of activities with slip danger; walking method ramp test*

A6.5　　　**LEGISLATION**

A6.5.1　　　*UK*

Acts of Parliament

Disability Discrimination Act 1995. 1995 c. 50 (ISBN 0-10-545095-2)

Health and Safety at Work etc Act 1974. 1974 c. 37 (ISBN 0-10-543774-3)

National Health Service and Community Care Act 1990. 1990 c. 19 (ISBN 0-10-541990-7)

Occupiers Liability Act 1957. 6 Eliz. 2 c. 31 (ISBN 0-10-850198-1)

Occupiers Liability Act 1984. 1984 c. 3 (ISBN 0-10-540384-9)

Regulations

The Building Regulations 2000

- Approved Document B. *Fire safety*. 2000 edition, consolidated with 2000 and 2002 amendments
- Approved Document K. *Protection from falling, collision and impact*. 1998 edition, amended 2000
- Approved Document M. *Access to and use of buildings*. 2004 edition

The Building (Scotland) Regulations 2004. SSI 2004/406

The Construction (Design and Management) Regulations 1994. SI 1994/3140 (ISBN 0-11-043845-0)

The Control of Noise at Work Regulations 2005. SI 2005/1643 (ISBN 0-11-072984-6)

The Control of Substances Hazardous to Health Regulations 2002. SI 2002/2677 (ISBN 0-11-042919-2)

The Food Safety (General Food Hygiene) Regulations 1995. SI 1995/1763 (ISBN 0-11-053227-9)

The Health and Safety (Consultation with Employees) Regulations 1996. SI 1996/1513 (ISBN 0-11-054839-6)

The Management of Health and Safety at Work Regulations 1999. SI 1999/3242 (ISBN 0-11-085625-2)

The Materials and Articles in Contact with Food Regulations 2005. SI 2005/898 (ISBN 0-11-072579-0) (supersedes 1987 Act)

The Provision and Use of Work Equipment Regulations 1998. SI 1998/2306 (ISBN 0-11-079599-7)

The Reporting of Injuries, Diseases and Dangerous Occurrences Regulations 1995. SI 1995/3163 (ISBN 0-11- 053751-3)

The Safety Representatives and Safety Committees Regulations 1977. SI 1977/500 (ISBN 0-11-070500-9)

The Workplace (Health, Safety and Welfare) Regulations 1992. SI 1992/3004.

A6.5.2   *EC*

Directives

The following European Community Directives are published by the Office of the Official Publications of the European Communities, Luxembourg. They are listed by Directive number. Where appropriate, the short title is given in parenthesis. OJ = *Official Journal of the European Communities.*

Council Directive 89/686/EEC of 21 Dec 1989 on the approximation of the laws of the member states relating to personal protective equipment ("the Personal Protective Equipment Directive"). *OJ* L399, pp 0018–0038

Council Directive 93/43/EEC of 14 Jun 1993 on the hygiene of foodstuffs ("the Food Hygiene Directive"). *OJ* L175, pp 0001–0011